城市复兴中的生活设施
Amenity in Urban Revival

建筑立场系列丛书 No.54

汉英对照
(韩语版第370期)

韩国C3出版公社 | 编

史虹涛 曹麟 时真妹 马莉 楚文潇 史瑞昕 陈帅甫 王单单 | 译

大连理工大学出版社

4

004 阿纳卡斯蒂亚渡口11街区大桥公园 _ OMA + OLIN

008 克里希-蒙费梅伊地铁站 _ Miralles Tagliabue EMBT + Bordas + Peiro

012 Vinge火车站 _ Henning Larsen Architects

016 匈牙利音乐厅 _ Sou Fujimoto Architects

020 布达佩斯照片博物馆与匈牙利建筑博物馆 _ Középülettervező Zrt

024 民族志博物馆 _ Vallet de Martinis Architectes + Diid Architectes

28 城市复兴中的生活设施

028 城市生活设施 _ Aldo Vanini

034 街道穹顶 _ CEBRA + Glifberg + Lykke

046 布什维克海滨公园 _ Kiss + Cathcart

054 湖滨的勒弗拉克中心 _ Tod Williams Billie Tsien Architects

064 欧什马戏艺术学院 _ Doazan + Hirschberger & Associés

074 Allez-Up攀岩中心 _ Smith Vigeant Architects

084 室内攀岩中心 _ Lanz + Mutschlechner + Wolfgang Meraner

城市住宅：

90 城市住宅-地域性-隐私性

090 城市住宅：城市住宅-地域性-隐私性

096 红房子 _ ISON Architects

112 SG住宅 _ Tuttiarchitetti

122 沙里夫哈住宅 _ Nextoffice

132 CM住宅 _ Bruno Vanbesien + Christophe Meersman

140 Namly住宅 _ Chang Architects

148 西原墙 _ Sabaoarch

156 津田沼住宅 _ Fuse-Atelier

166 1014住宅 _ Harquitectes

182 建筑师索引

C3 建筑立场系列丛书 No.54

4

004 Anacostia Crossing, 11th St. Bridge Park _ OMA + OLIN

008 Clichy–Montfermeil Metro Station _ Miralles Tagliabue EMBT + Bordas + Peiro

012 Vinge Train Station _ Henning Larsen Architects

016 The House of Hungarian Music _ Sou Fujimoto Architects

020 Photo Museum of Budapest and Museum of Hungarian Architecture _ Középülettervező Zrt

024 Museum of Ethnography _ Vallet de Martinis Architectes + Diid Architectes

28 Amenity in Urban Revival

028 *Urban Amenity _ Aldo Vanini*

034 Streetdome _ CEBRA + Glifberg + Lykke

046 Bushwick Inlet Park _ Kiss + Cathcart

054 LeFrak Center at Lakeside _ Tod Williams Billie Tsien Architects

064 Circus Arts Conservatory _ Doazan + Hirschberger & Associés

074 Allez-Up Climbing Center _ Smith Vigeant Architects

084 Indoor Rock Climbing Center _ Lanz + Mutschlechner + Wolfgang Meraner

90

Urban Dwell
Proxemics - Territoriality - Privacy

090 *Urban Dwell: Proxemics-Territoriality-Privacy _ Fabrizio Aimar*

096 Red House _ ISON Architects

112 SG House _ Tuttiarchitetti

122 Sharifi-ha House _ Nextoffice

132 House CM _ Bruno Vanbesien + Christophe Meersman

140 Namly House _ Chang Architects

148 The Wall of Nishihara _ Sabaoarch

156 House in Tsudanuma _ Fuse-Atelier

166 House 1014 _ Harquitectes

182 Index

城市连接 Urban Link

阿纳卡斯蒂亚渡口11街区大桥公园 _OMA+OLIN

OMA联手景观设计师OLIN赢得了竞赛,成为华盛顿特区11街区大桥公园的设计团队。他们提出的11街区大桥公园设计方案是"交流之地"。历史上,阿纳卡斯蒂亚河的两岸从无交集,而今,这座横跨河上的公园将利用一系列户外项目空间和活动区域来连接两岸,这些区域提供一处既悬浮于河流又固定于两岸的迷人区域。

为了实现这一方案,设计团队将大桥公园设计为一个明确的交流沟通的契机,河流两岸在这里交汇共存。阿纳卡斯蒂亚河大桥公园将提供一个多层次的项目方案,向人们呈现一处全新的街区公园,一处附近工作的人们在业余时间的休闲去处,一处居民疗养之所,以及一处等待游客开发的观光场所。从河两岸延伸而来的道路是大跳板一样的斜坡,将游客们带到高处的最佳位置观赏两侧的地标性景观。阿纳卡斯蒂亚桥面上的路在河面上延伸,并形成环路,环绕从海军基地延伸出来的小径,并以一个独特的姿态连接了河对岸。桥梁的最终形态形成了一种标志性的交汇,即X形,成为河岸的全新形象,很容易被人们所识别。这座桥梁成为一个独特且标志性的结构,其特色和根本致力于使社区与河岸景观相通。通过规划的活动,桥梁展现出该地域独特的文化和自然历史。为了鼓励游客常年来到大桥以及附近的社区,沿大桥一侧设置了休闲和餐饮设施、极端气候应对设备,以及各种季节性的功能规划。此外,桥梁提供了参加附近社区的各种活动的入口。

两条道路在桥面交汇,形成一个中央汇聚点——一个开放的广场。这个开放广场能全年提供一个用作市场,或者举办节日活动和戏剧的灵活场所。环绕着广场四周的小路进一步突显了桥梁作为集娱乐、休闲、学习和机会为一体的活动枢纽的特性。

Anacostia Crossing, 11th St. Bridge Park

OMA won the competition for the 11th Street Bridge Park in Washington DC with landscape architect OLIN. The design for the 11th Street Bridge Park is a place of exchange. The park at Anacostia Crossing will connect two historically disparate sides of the river with a series of outdoor programmed spaces and active zones that will provide an engaging place hovering above, yet anchored in, the Anacostia River. To create this place the team has designed the bridge park as a clear moment of intersection where two sides of the river converge and coexist. Anacostia Crossing will offer layered programs, presenting a new neighborhood park, an after-hours destination for the nearby workforce, a retreat for residents and a territory for tourists to explore. Paths from each side of the river operate as springboards – sloped ramps that

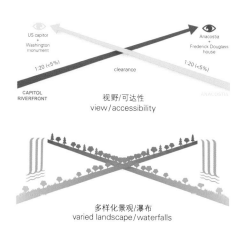

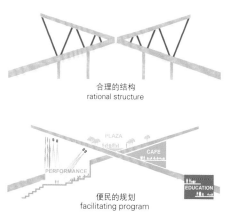

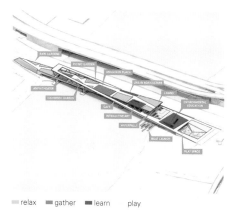

1. Rain Garden
2. Capitol Hill Look Out
3. Amphitheater
4. Hammock Garden
5. Picnic Garden
6. Fern Garden
7. Anaquash Plaza
8. Urban Agriculture
9. Lawn
10. Interactive Art
11. Sculpture Lawn
12. Anacostia Look Out
13. Kayak & Paddle Boat Launches
14. Waterfall
15. Environmental Education Center
16. Play Space
17. Anacostia Park

elevate visitors to maximized look-out points to landmarks in either direction. Extending over the river, the Anacostia paths join to form a loop, embracing the path from the Navy Yard side and linking the opposing banks in a single gesture. The resulting form of the bridge creates an iconic encounter, an "X" instantly recognizable as a new image for the river. While the bridge is a unique and iconic structure, its character and essence are rooted in making this river landscape accessible to the community. Through programmed activities the bridge will showcase the region's unique cultural and natural history. To encourage visitors to spend time on the bridge and neighboring communities throughout the year, amenities for comfort and refreshment, mitigation of climate extremes, and opportunities for seasonal programming are provided along the entire length of the bridge. The bridge provides a gateway to events with strong roots in the adjacent communities.

The intersection point of the two paths shapes the central meeting point of the bridge – an open plaza that provides a flexible venue for markets, festivals, and theatrical performances held throughout the year. The paths that frame this plaza further enhance the bridge as a hub of activity, providing a sequence of zones designated for play, relaxation, learning and gathering.

克里希-蒙费梅伊地铁站 _Miralles Tagliabue EMBT + Bordas + Peiro

大巴黎协会宣布Miralles Tagliabue EMBT和Bordas+Peiro在设计未来法国巴黎克里希-蒙费梅伊地铁站的大赛中获胜。

大巴黎计划是一项宏伟的项目,旨在使现有的运输网络现代化,并且在巴黎大都会建造一条名为"大巴黎快线"的全新自动化地铁线。克里希-蒙费梅伊社区曾在2005年10月到11月经历了暴动,车辆和公共建筑都有所损毁。大巴黎快线在该地区建造的新地铁站将成为展示该地区变化的一个标志。

新地铁站位于两座市郊小镇克里希和蒙费梅伊的边界,将成为连接地铁、新有轨电车和公交车的重要枢纽。当下,社区正在经历城市化历程,并且初步成效已经显现(新的公寓社区),但广场周围的空间依然百废待兴。一条绿色连接通道将穿过这个区域(追随了古老的"Dhuis沟渠"的路径),并且成为新广场的重要组成部分。

这种设计方式将地铁站通道与城市背景进行了最大程度的融合。地铁站通道是一条平缓的楼梯斜坡,邀请走出地铁站的人们进入广场。蔓藤架式的屋顶覆盖了入口、地铁站的设施和建筑、自行车停车场,并且一直延伸到广场的重要区域,保护着将成为市场的区域。

地铁站的理念是在考虑到该地区众多居民起源的基础上,给予这个地区一个新的身份。建筑师意图把一处灰色废弃的地区变成活跃的彩色广场,使人们得到欢愉,变得乐观。这就是建筑师选择将铺路、屋顶的形状与颜色、装饰性主题与非洲的色彩作为主要设计元素的原因。

当你进入地铁站时,进站和出站的人流将汇聚在一处简单的特殊空间,该空间透过天窗的自然光增添了空间的活力。通道在此成为一个游戏,乘客们能看到对方但却不会相遇。就像一场戏剧,乘客们被带到了舞台上,成为主角。覆盖在上空空间顶部的有机形状的面板赋予其独特的空间特性。

Clichy-Montfermeil Metro Station

Miralles Tagliabue EMBT and Bordas+Peiro were announced as winners by the Société du Grand París, to design the future metro station Clichy-Montfermeil in Paris, France.
The Grand Paris is an ambitious project to modernize the existing transport network and create a new automatic metro – the Grand Paris Express for the Paris metropolitan area. This neighborhood was the scene of the violent riots that occurred in October and November 2005 when cars and public buildings were burnt. A new station of the Grand Paris Express will be a symbol of the change for these areas. The new station is located at the border of the two suburb small towns: Clichy-sous-bois and Montfermeil and is an important pole of connection of the metro network with new tram line and bus lines. In present, the neighborhood is engaged in a major urban operation whose first result can already be seen(new blocks of flats), but the space around the square is unstructured and abandoned. An important green connection crosses the place following the course of the ancient "Aque-

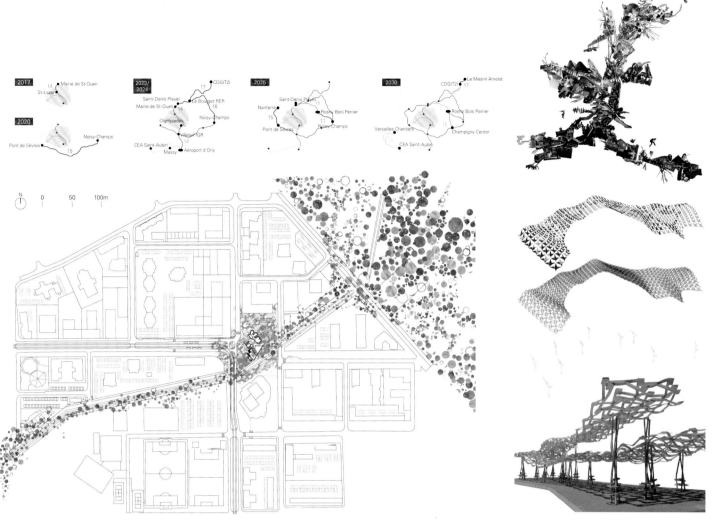

duct de la Dhuis" and this connection will be part of the new square.

The approach is looking for the maximum integration of the metro station's access into the urban context. The access is made through a slow slope with stairs that invites the people coming out of the station to participate on the square. A pergola-roof is covering the entrance, the installations, the building of the station, and the bicycle parking and continues on an important part of the square, protecting an area dedicated in the future to a periodical market.

The idea was to give a new identity to this place, with a glance to the origins of many of his inhabitants. The architects wanted to transform this gray and abandoned place into a vivid and colorful square, which inspires joy and optimism. This is why they are based on the motives of the pavement, the shapes and the colors of the roof on the tissues, decorative motives and colors from Africa.

Once entering the metro station, the circulation of the passengers towards and from the platforms is organized in a single unique space receiving natural light from a skylight. The access becomes a game, where the passengers can see each other but they don't cross. The circulation has been brought into scene, and the passenger becomes the principal actor. The character of this unique space is given by the panels that cover the organic shape of the void.

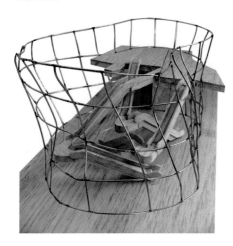

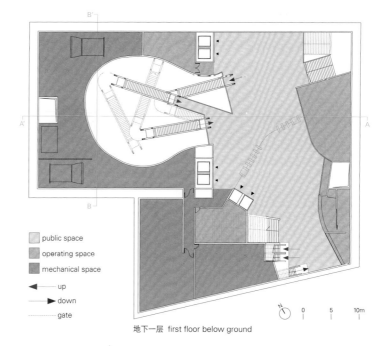

地下一层 first floor below ground

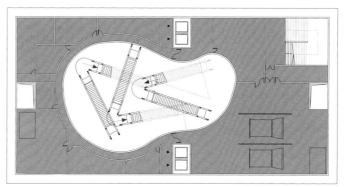

地下二层 second floor below ground

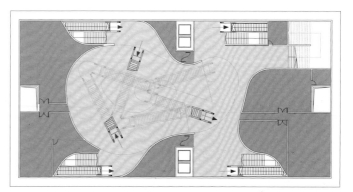

地下三层 third floor below ground

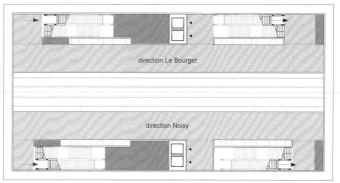

地下四层 fourth floor below ground

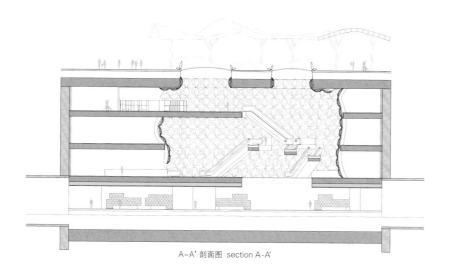
A-A'剖面图 section A-A'

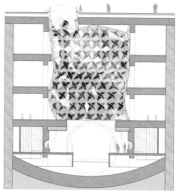
B-B'剖面图 section B-B'

Vinge火车站 _Henning Larsen Architects

Henning Larsen建筑师事务所近期赢得了腓特烈松市一个地区性火车站和一个新公共广场的项目竞标，该项目位于丹麦首都哥本哈根北部未来计划建造的Vinge城镇。

Vinge城镇占地350公顷，它将是丹麦最大的城市建设项目。在新城市规划图的中间，一个圆形火车站与周围环境有机地融合在一起。火车站的城市空间和景观延伸并且交汇，以跨越轨道，确保铁轨不会将城镇分成两部分。

Vinge火车站被设计为城镇开发的中心，并整合景观和城市的动态性。火车站非定向的椭圆形状将周围景观结合在一起，起伏的地形创造出了一个平静的中心。

火车站位于城镇中心，它为公众提供了方便的公共交通通道。这是Vinge城镇的可持续性方面之一。它鼓励更多的人乘火车，而不是开车上班和上学。

火车站和绿化中心不仅仅作为桥梁，它们还都置于和轨道相同的平面内，在视觉上和物理上连接了两个水平线。火车站下部区域还建造了很多覆顶的铁路站台和商店。

在Vinge城镇中，多样性和可持续性将成为城市的综合开发核心。

Vinge Train Station

Henning Larsen Architects has won Frederikssund municipality's architecture competition to design a regional train station and new quarter in the future town of Vinge, the north of Copenhagen, Denmark.

Covering 350 hectares, it will be the largest urban development project in Denmark. In the middle of the new town plan, a circular station adapts organically to its surroundings. The station's urban space and the landscape stretch and meet to span the rails, ensuring that the railway does not divide the town into two parts. Vinge Train Station has been designed to function as the heart of the development,

and to unify the movements of landscape and city. The station's undulating topography creates a calm center, as the non-directional elliptical shape brings the surroundings together.

Located in the center of the city, the train station offers convenient access to public transportation. This focus forms one of the many sustainable aspects of Vinge, as more people will be encouraged to take the train to work and go to school as opposed to going by car.

Instead of merely functioning as a bridge, the station and the green heart are placed at the same level as the rails, visually and physically connecting the two levels. A space under the station is thereby created with the covered train platforms and shops.

Vinge will become a city where diversity and sustainability are essential to its comprehensive development.

A-A' 剖面图 section A-A'

B-B' 剖面图 section B-B'

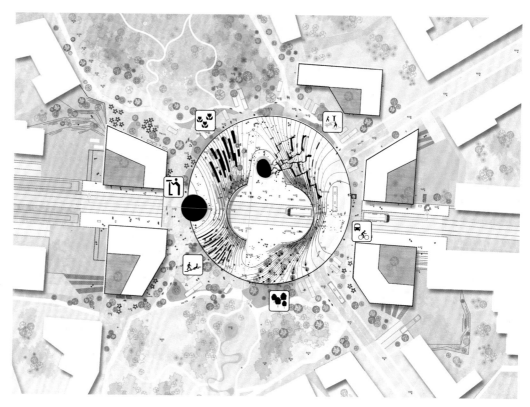

布达佩斯公园 Budapest Park
匈牙利音乐厅 Sou Fujimoto Architects

位于布达佩斯城市公园中心的匈牙利音乐厅不仅仅是一座博物馆,还是一个涵盖了过去与将来、人类与文化、自然和音乐科学的场所。它的功能并非是陈列,而是邀请人们进入其中,促使人们对音乐的态度从简单的沉思转入更深内涵的参与和互动。在这里,听,即声音和音乐的本质,可以从更深层次的角度得到体现。与传统的展馆、教室或者活动大厅不同,建筑师们选择了由建筑引导游客这一设计技巧。厅内和音乐厅外的公园类似,没有明确的人行通道,人群如同随性的河流一般肆意涌入音乐厅,好似声音穿透过空间,在表面弹跳,沿着墙壁奔跑。

匈牙利音乐厅也是21世纪博物馆的一种展望,利用各种方式高度融于环境中。无论从生态角度还是美学角度,都细致地融入周边景观内和公园的灵魂中。这一对21世纪的展望将焦点放在相遇和分享方面,无论你是来学习、聆听音乐、表演、演奏或工作,每个人都能在这里教与学。

这个项目成为象征着当今世界上旅游和通讯方面无边界且自由电子化的一个标志。

这个项目外形的灵感来源于自然生长的树冠。圆形的体量轻轻浮起,向四周延伸,释放出底层的空间来欢迎从四面八方而来的客人。建筑顶部的穿孔能让自然光通过,就像阳光透过森林树木的叶子间那般。

诸如音乐会等丰富多彩的活动在漂浮的体量的下方进行,使人们可在此相遇并倾听音乐,由此吸引大批观众在此会面并且分享各自的音乐。开阔的一层使室内和室外的界限消失,且成为周边景观的延续,使绿化区域如声波一样向周围延伸。游客在林间的博物馆内无拘无束,沿着曲形的墙壁,在宽敞的螺旋状楼梯上下行走,沉醉于周围的空间以及柔和且变化的日光中。路上所遇的各种惊喜,就仿佛曲子中那些出乎意料的音符一样,遵循上下内外不可打断的步调移动,将博物馆、公园、游客、音乐融于一体,为游客带来一段独一无二的享受。

The House of Hungarian Music

The House of Hungarian Music, in the heart of the City Park of Budapest, is not only a museum but a larger vision encompassing past and future, people and culture, nature and the science of music. It is not about displaying but inviting in, going away from the simple contemplation to suggesting participation and interactions. The essence of sound and music, hearing, can be brought in much deeper ways. Away from the conventional exhibition spaces, classrooms or event halls, the architects chose to let architecture cradle the visitors along their way. No clear path, pedestrians meander around the mu-

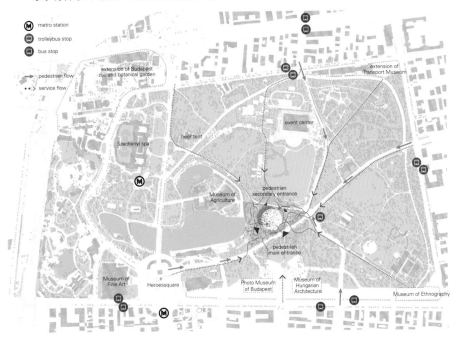

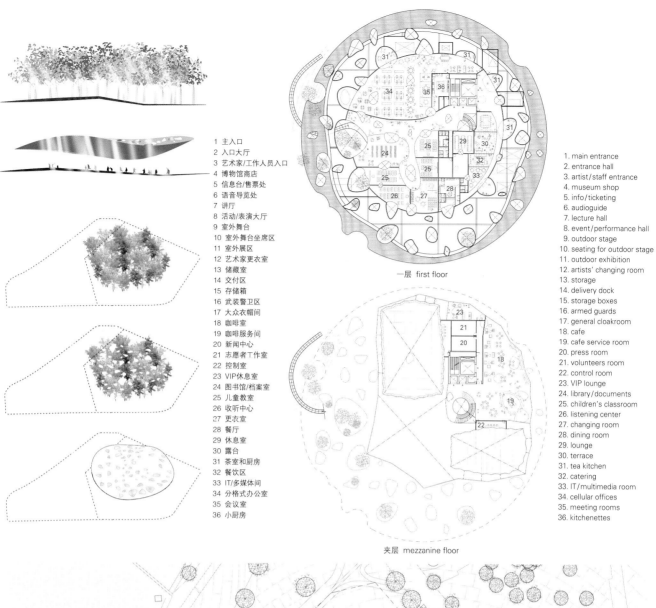

1 主入口	1. main entrance
2 入口大厅	2. entrance hall
3 艺术家/工作人员入口	3. artist/staff entrance
4 博物馆商店	4. museum shop
5 信息台/售票处	5. info/ticketing
6 语音导览处	6. audioguide
7 讲厅	7. lecture hall
8 活动/表演大厅	8. event/performance hall
9 室外舞台	9. outdoor stage
10 室外舞台坐席区	10. seating for outdoor stage
11 室外展区	11. outdoor exhibition
12 艺术家更衣室	12. artists' changing room
13 储藏室	13. storage
14 交付区	14. delivery dock
15 存储箱	15. storage boxes
16 武装警卫区	16. armed guards
17 大众衣帽间	17. general cloakroom
18 咖啡室	18. cafe
19 咖啡服务间	19. cafe service room
20 新闻中心	20. press room
21 志愿者工作室	21. volunteers room
22 控制室	22. control room
23 VIP休息室	23. VIP lounge
24 图书馆/档案室	24. library/documents
25 儿童教室	25. children's classroom
26 收听中心	26. listening center
27 更衣室	27. changing room
28 餐厅	28. dining room
29 休息室	29. lounge
30 露台	30. terrace
31 茶室和厨房	31. tea kitchen
32 餐饮区	32. catering
33 IT/多媒体间	33. IT/multimedia room
34 分格式办公室	34. cellular offices
35 会议室	35. meeting rooms
36 小厨房	36. kitchenettes

一层 first floor

夹层 mezzanine floor

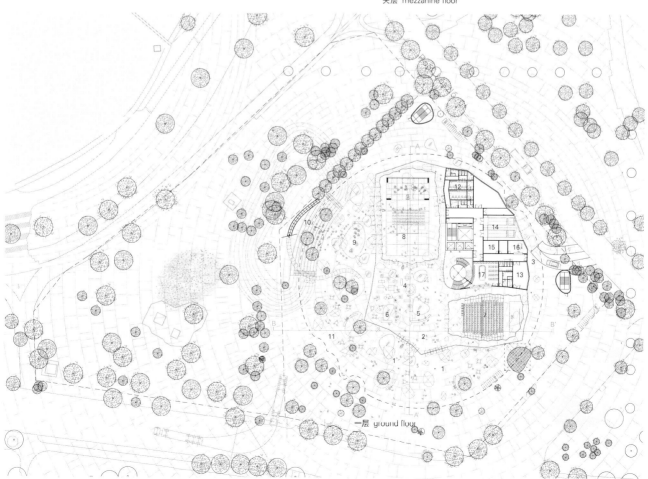

一层 ground floor

1 展览空间	1. exhibition space
2 圆顶式空间	2. dome
3 入口大厅	3. entrance hall
4 讲厅	4. lecture hall
5 冷却装置	5. chiller
6 机械区	6. mechanical area
7 收听中心	7. listening center
8 IT/多媒体间	8. IT/multimedia room
9 更衣室	9. changing room
10 控制室	10. control room

A-A' 剖面图 section A-A'

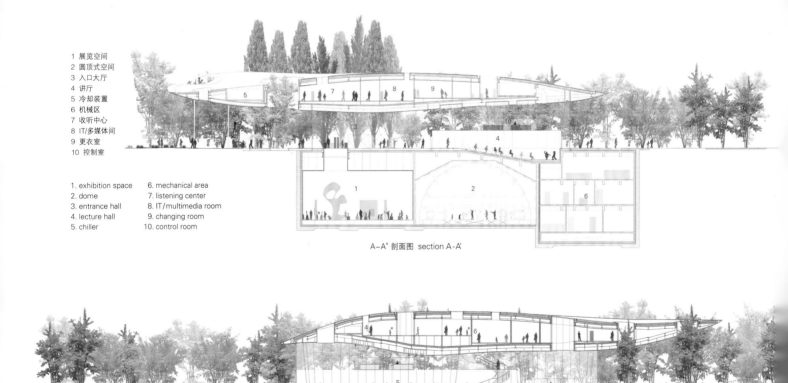

1 临时展区	1. temporary exhibition
2 可移动的墙体	2. mobile wall
3 永久性展区	3. permanent exhibition
4 办公室	4. office
5 活动大厅	5. event hall
6 图书馆	6. library

B-B' 剖面图 section B-B'

seum, as they would do around the park, invading the space like a continuous flow, just as sound permeates space, bouncing around surfaces, running along the walls. The House of Hungarian Music is also the vision of a 21st century museum, highly integrated into the environment by all means. Ecologically friendly as well as aesthetically, fitting carefully to the landscape and the soul of the park, this vision of the 21st century puts the emphasis on the act of meeting and sharing, regardless of the reasons that brought people to this place: to study, listen, perform, play or work, everyone has a place to teach and to learn. The project becomes a symbol of today's open borders and electronic freedom of travelling and communicating all around the world.

The project finds its shape naturally among the trees' crowns. Its circular volume levitates gently, turning to every direction, and freeing out the ground floor to welcome people from all around, while its perforations let the natural light through, just like the sun rays breaching between the leaves of a forest.

Activities such as concerts or various events take place under the floating volume, for everyone to see and listen, attracting large crowds to meet and share music. This open ground-floor, where limits between inside and outside are fading away, is thought as the continuity of the landscape whose green areas radiate away, as the waves of sound. A visitor wanders freely among the museum, between the trees, along the curves of the walls, up and down the large spiral staircase, cradled around by the vibrations of the spaces, and by the soft variations of the sunlight. Surprising encounters punctuate one's route, just as the unexpected notes of a melody, following an uninterrupted movement, up, down, around, inside, outside, the same flow binding softly together museum, park, people and music, making the visitor's experience unique.

布达佩斯照片博物馆与匈牙利建筑博物馆

Középülettervező Zrt

拟建博物馆建筑的位置、形状以及外观与重新构建的Rondo公园有着密切的联系。这座建筑置于此处，其在建筑方面的特点是通过调整以符合公园的规模。两个形式严格的方形体量将Rondo街道封闭起来，转而向城市延伸，使其与滨海大道和Dózsa György út车站平行。这种姿态通过微小的建筑的转动得以突出。两座建筑物间的不起眼的广场同时有着不同的功能：1956纪念场所，两座博物馆间的城市广场，城市与公园的连接通道，同时也是各种城市活动的举办场所。广场以Rondo中心的1.5公顷的草地为背景，树木遮挡了通向公园的视野。博物馆设有多个楼层，使其能够实现如下功能：展出任务，遵循公园最少的"碳足迹"设计概念，以及做出最重要的创造性举动——营造出一种标志性的城市气息。两座建筑的外形与规模几乎一致。功能差异则体现在展出空间的结构和立面的细节设计方面。

匈牙利建筑博物馆的构造体量，以及双高空间增建的室内结构和连锁结构展现了建筑的戏剧性。预制混凝土立面上的碎片式版画展现了博物馆的主题。

布达佩斯照片博物馆的双层立面结构，位于透明的外部玻璃墙与暗色LED媒体墙之间的楼梯和流线都参考了光学艺术。建筑外部这种"外向"的表现，将使建筑和周围环境保持永恒的交流和连通。

Photo Museum of Budapest and Museum of Hungarian Architecture

The location, the form and the appearance of the proposed museum buildings are in close interaction with the reformulated Rondo. The architectural character of the buildings to be placed here is adjusted to the Park's scale. The two strictly formed square blocks are closing the alley of the Rondo into the direction of the city while turning it to the Promenade parallel with Dózsa György út. This gesture is enhanced with a minor rotation of the building masses. The plain Square between the two buildings has multiple functions at the same time: venue of the 1956 commemoration, urban square between two museum buildings, a link between the city and the park and also a space for urban events. The 1.5 hectares grassy meadow at the middle of the Rondo gives a background to the square while the trees are closing the view towards the park. The multi-level arrangement of the museums is functionally justified by the

exhibition brief, the principle of the smallest "footprint" in the Park and last but not least the creation of a characteristic urban accent. The design geometry and basic dimensions of the two buildings are nearly the same. The difference in the functional content is expressed by the exhibition space structure and the design details of the facades.

The tectonic mass of the Hungarian Museum of Architecture, the additive construction of the interiors and the interlocking structures of the double-height spaces reveal the drama of architecture. The fragmental graphic engravings on the prefabricated concrete facade elements are reflecting the main theme of the museum.

The double-layered facade construction of the FotoMuzeum Budapest, the stairs and circulation located between the transparent exterior glass wall and the dark colored LED media wall refer to the art of Light. This "extroverted" behavior of the building exteriors will keep a permanent communication and interaction with the surroundings.

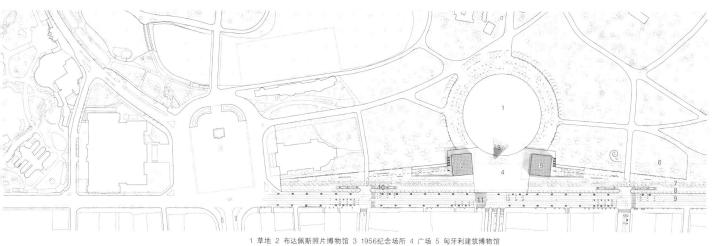

1 草地 2 布达佩斯照片博物馆 3 1956纪念场所 4 广场 5 匈牙利建筑博物馆
6 根据区域规划的地下停车库 7 人行道 8 自行车车道 9 Dózsa György út车站 10 通往地下停车库的斜坡 11 人行十字路口
1. the meadow 2. Photo Museum of Budapest 3. 1956 memorial 4. square 5. Museum of Hungarian Architecture
6. underground garage according to zoning plan 7. pedestrian 8. bicycle route 9. Dózsa György út station 10. ramps to underground garage 11. pedestrian crossing

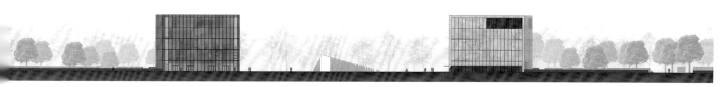

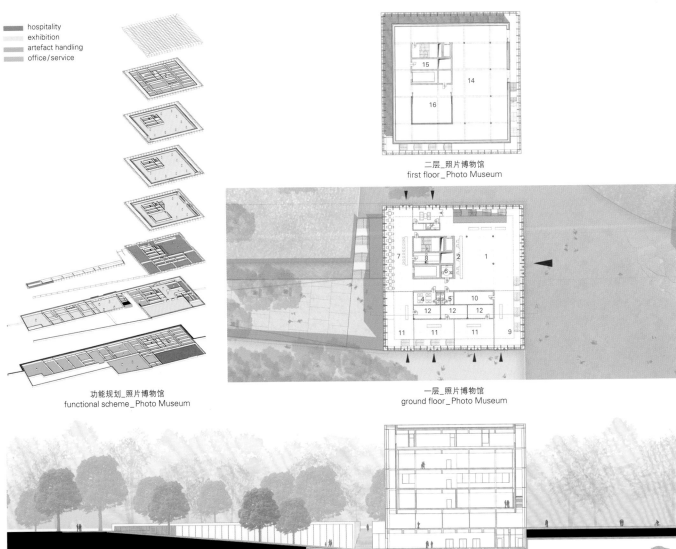

hospitality
exhibition
artefact handling
office/service

功能规划_照片博物馆
functional scheme_Photo Museum

二层_照片博物馆
first floor_Photo Museum

一层_照片博物馆
ground floor_Photo Museum

A-A' 剖面图_照片博物馆
section A-A'_Photo Museum

1 入口大厅	9 博物馆商店
2 售票处	10 商店存储处
3 VIP休息室	11 其他的商店
4 志愿者办公室	12 其他商店存储处
5 急救室	13 影像室
6 婴儿护理室	14 展览品存储处
7 咖啡室	15 流线
8 咖啡服务室	16 放映室
1. entrance hall	9. museum shop
2. info tickets	10. shop storage
3. VIP lounge	11. other shop
4. volunteer's room	12. other shop storage
5. first aid room	13. video room
6. baby care	14. exhibition storage
7. cafe	15. circulation
8. cafe service room	16. project room

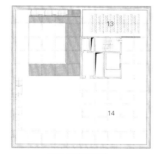

二层_建筑博物馆
first floor _ Museum of Architecture

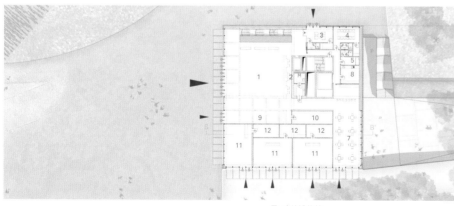

一层_建筑博物馆
ground floor _ Museum of Architecture

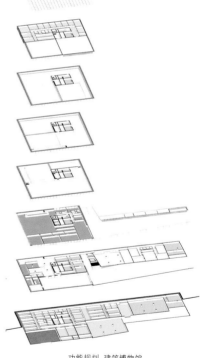

功能规划_建筑博物馆
functional scheme _ Museum of Architecture

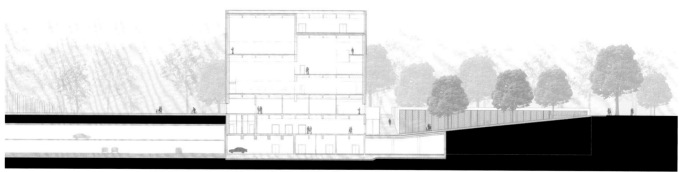

B-B' 剖面图_建筑博物馆
section B-B' _ Museum of Architecture

民族志博物馆 _Vallet de Martinis Architectes + Diid Architectes

民族志博物馆是全国级的文化项目，从城市规划角度来看将产生重大的利益。建筑使市民聚集并居住在博物馆所在地。公众来去自由，与空间建立有意义的联系。建筑的各个部分都为这种互动提供方便，并使之持续下去。它将成为受人欢迎的会面场所，也是探索民族志、城市和自身的起点。

团队在建筑下方设计了一处宽敞的开放式广场，来扩大建筑的公共空间。

建筑不受前方实际边界（立面的边界）的限制，而是和周围环境融为一体。空间中心的公共广场活力十足，与众不同，很有吸引力。

该建筑结构合理，形态简单，紧凑，气势宏伟，并且形式上不张扬，实用，质朴，简约，既与城市相融又与众不同。

大楼的建筑结构与城市肌理统一。博物馆在城市景观中非常醒目，热情邀请城市和市民来访。团队旨在设计一座能同时和城市、地区和相邻公园进行互动的建筑。

作为Dózsa György út大道上独特的城市地标，博物馆是当地文化繁荣发展的重要部分。博物馆为当地社区提供新的文化设施，公众可以随时使用。通过创造接纳和吸引公众的空间，内部功能保持和外部景观的贯通连接。

博物馆西南侧新开设的出入口可以与城市公园相连。

Museum of Ethnography

This cultural project will prove to be nationwide in scope, and of significant interest from an urban planning perspective. The building will enable citizens to gather at and inhabit the museum's site. The public will be able to move freely and develop a meaningful connection with the space. Each part of the building will facilitate this interaction and avoid any suggestion of finality, dead-ends. It will instead become a welcoming meeting place, and thus a departure point from which to explore Ethnography, the city and the self.

The teams chose to expand the building's public space through laying-out a generous open square under the building.

The structure would no longer be limited to its actual borders – the perimeter of its facade. It instead integrates with and gets caught up in the surrounding environment. A dynamism and a certain gravitational pull, at the center of this space works to create a distinct public square. It is a rational structure with a simple morphology, compact and monolithic. The architecture aims for a moderate form, functional, sober and minimal in character. It is integrated into the city and stands out in its locale.

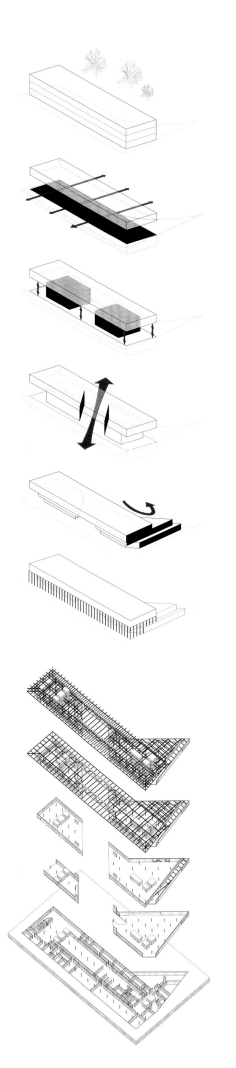

1 技术室
2 办公室
3 会议室

1. technical room
2. offices
3. meeting room

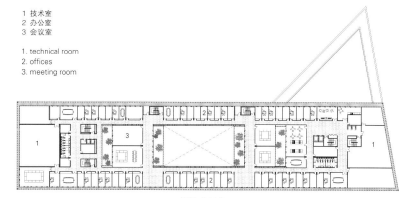

四层 third floor

1 儿童博物馆　　4 博物馆自习室
2 讲室　　　　　5 永久展厅
3 活动厅　　　　6 露台

1. children museum　　4. museum learning
2. lecture hall　　　　5. permanent exhibition
3. event hall　　　　　6. terrace

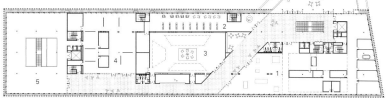

三层 second floor

1 广场
2 公共入口　　　5 博物馆商店
3 衣帽间　　　　6 自助餐厅
4 售票处　　　　7 永久展厅

1. square
2. public entrance　　5. museum shop
3. cloakroom　　　　　6. cafeteria
4. info ticketing　　　7. permanent exhibition

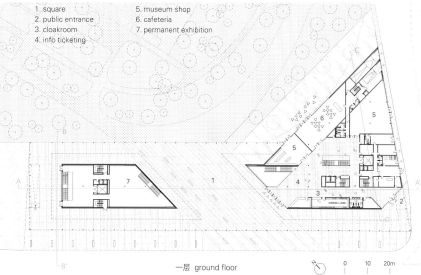

一层 ground floor

The building is in unity with the city. The museum marks its presence in the urban landscape while warmly inviting the city and its citizens inside. The teams sought to design a building which interacts simultaneously with the city, the local neighborhood and the neighboring park.
The museum forms a distinctive city landmark along Dózsa György út boulevard, proving an integral part of the neighborhood's evolving cultural rejuvenation.

For the local neighborhood the museum would bring new cultural amenities, while remaining accessible to the public. The interior forges a permanent link with the exterior landscape by creating spaces which are welcoming and entice the public inside.
The museum also connects with the Varosliget park by providing a new gateway to its south-western side.

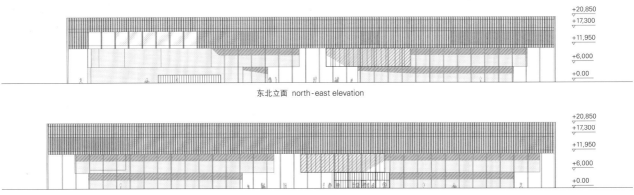

东北立面 north-east elevation

西南立面 south-west elevation

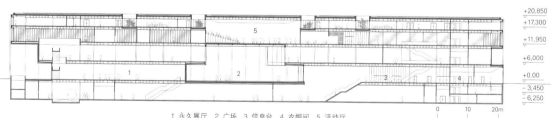

1 永久展厅 2 广场 3 信息台 4 衣帽间 5 活动厅
1. permanent exhibition 2. square 3. information 4. cloakroom 5. event hall
A-A' 剖面图 section A-A'

1 永久展厅 2 技术室
1. permanent exhibition 2. technical room
B-B' 剖面图 section B-B'

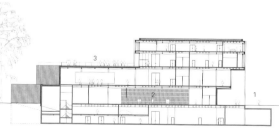

1 公共入口 2 博物馆商店 3 露台
1. public entrance 2. museum shop 3. terrace
C-C' 剖面图 section C-C'

城市复兴中的生活设施
Amenity in Urban

人们不再为了生计终日劳作,时间越来越富余,使人们投身于非功利性的活动成为可能。实现现代化后,社会重新发现体育运动的价值,不论是竞技性运动还是休闲运动。希腊和罗马文明推崇运动,但在随后基督教占主导的世纪里,体育运动又逐渐衰落。

我们的社会对竞技性体育项目投入了大量的财力,进行了广泛的报道,并兴建了功能明确的专用场馆。而休闲体育运动和非竞技性体育运动场馆的设计就不那么新颖,资源投入也少,且仅作为教学用场所使用。

如果现代竞技性体育终将陷入娱乐业范畴,成为生活中的上层建筑,请别忘了"sport"一词来源于拉丁语"deportare",意为"出去",也来源于古法语词"desport",意为"消遣,娱乐"。

当代建筑文化的缺陷之一就是建筑功能难以与时俱进,用途单一,喜欢追逐一时的潮流。

事实上,休闲运动形式多样,不拘一格,尤其是失去原用途的或已被纳入城市发展却没有明确功能定位的城市空间的改建,为建筑设计和功能研究提供了用武之地。在修建休闲运动和非竞技性体育运动的场所和设施的过程中,人们可以尝试用新形式和新方法,把传统城区连接起来,重返那些失去原有功能的地区。

The gradual freeing of human time from the totalitarian demands of survival has expanded the possibility, if not the necessity, of humanity's dedication to non-utilitarian activities. Only with the advent of modernity has society rediscovered the values of sport, both competitive and recreational, embraced by the Greek and Roman civilizations, which seemed to disappear from popular diffusion in later centuries dominated by Christian ethics.

Our society invests substantial economic resources and media attention to the practice of competitive sport, and has developed for this purpose specific, well-defined architectural typologies. Less creativity and resources have been allocated to places of leisure and non-competitive physical activity, with such efforts often limited to the definition of spaces directly connected to educational complexes.

If competitive sport has now finally slipped into the field of show business, playing a superstructural role in everyone's lives, we cannot forget that the word "sport" originates from the Latin "deportare", in the sense of "get out" and then from the old French word desport, meaning recreation, fun.

Yet one of the limitations of contemporary architectural culture consists in its difficulty imagining new functions to be facilitated using the means of the discipline, while it instead wearily performs a repertoire that varies little and that always favors transient, self-referential fashion.

In fact, leisure's wide variety of activities and non-conventional character make it an interesting field of compositional and functional research, particularly as regards the redevelopment of urban spaces that have lost their original purpose or that have been incorporated into urban growth without a specific function. Sites and facilities related to leisure and non-competitive physical exercise represent a ground for experimentation in new forms as well as in finding new ways to connect conventional parts of the city or to re-enter areas that have lost their original function.

Revival

街道穹顶_Streetdome/CEBRA+Glifberg+Lykke
布什维克海滨公园_Bushwick Inlet Park/Kiss+Cathcart
湖滨的勒弗拉克中心_LeFrak Center at Lakeside/Tod Williams Billie Tsien Architects
欧什马戏艺术学院_Circus Arts Conservatory/Doazan+Hirschberger & Associés
Allez-Up攀岩中心_Allez-Up Climbing Center/Smith Vigeant Architects
室内攀岩中心_Indoor Rock Climbing Center/Lanz+Mutschlechner+Wolfgang Meraner
城市生活设施_Urban Amenity/Aldo Vanini

城市生活设施

如今,无论是竞技性体育运动还是休闲性体育运动,在我们社会中都占有重要位置,然而100多年前却没有专门的运动场,这确实很有意思。事实上,希腊和罗马之后,也许是受到基督教道德规范和基督教对肉体束缚教义的影响,人们似乎抛弃了不实用体育活动带来的乐趣。像佛罗伦萨足球这样的休闲体育运动被贵族保留下来,进行初级战术训练,偶尔在显示社会凝聚力或宣泄情感的特殊场合进行活动。

随着实证主义和社会科学的兴起,休闲体育运动的基本社会角色得以恢复。19世纪末奥林匹克理想理念的回归,也确立了竞技性体育的基本社会角色。竞技性体育活动成为媒体的宠儿,不仅恢复了竞技性体育运动的主宰地位,使之成为最赚钱的经济行业,还催生了一大批现代建筑类型,从体育场到健身馆,从游泳池到大型运动中心,处处显示出这个时代的恢宏气势。这些场馆在媒体的宣传下,传播至世界各地,也成了建筑明星大腕们展示的舞台。

公众越来越喜欢观看职业运动,同时,自发的体育活动,如街头运动,也越来越受欢迎。这些在不完善的城市环境中自发形成的运动,现在需要特定的设施,来提供优质的服务,使民众有机会聚集在一起,为此,那些可以为周围环境带来新价值的生活设施就成为有效的解决办法。

社会间的互动由此快速发展,原来简单的家和单位两点一线的城市模式开始变得复杂。人们在基本的居家和工作以外获得更多的空闲时间。相应地,人们也需要利用这些时间,来减缓繁杂的日常生活带来的压力。

同时,从工业社会过渡到后工业社会,大都市扩张,一些工厂要搬

Urban Amenity

Given how sporting activities, both competitive and recreational, now command center stage in our society, it is interesting to note that just over a century ago there were no places or buildings designated for such activities. It seems, in fact, that after the Greek and Roman experience, humanity renounced the pleasures of non-utilitarian physical activity, probably under the influence of the moral dictates of Christianity and its wariness of every aspect of corporeality. Recreational physical activities were reserved for aristocratic classes as a form of initiation into the art of war, or for rare, special moments of social cohesion and liberating outbursts, such as in Florentine football.

The fundamental social role of recreational physical activity would be rediscovered only with the advent of positivism and the social sciences, and that of competitive sports with the return of the Olympic ideal at the end of the XIX century. The increasing media popularity of competitive sports, in addition to restoring their status as a formidable instrument of consent and turning them into one of the most profitable economic sectors, has occasioned the development of a multitude of modern architectural typologies, ranging from stadiums to gyms and from swimming pools to large sports halls, all scenes of the celebration of this era's social splendors. Places and spaces amplified by the power of the media into a global dimension have now become the playground of the most celebrated archistars.

Parallel to the steadfast growth in the public's interest in the spectacle of professional sports, unorganised physical activity, such as street sports, has also become increasingly popular. Spontaneously born within sketchy urban situations, such activities now require specific facilities that interpret the role of significant services and offer opportunities for social aggregation. Solutions in this regard represent amenities that can bring new value to a context.

The rapid evolution of social dynamics has moved from an urban model based on a simple duality of residence and production to a complex society in which individuals, moving beyond elementary dwelling and working, have access to more free time. With this availability, however, comes a corresponding necessity to make use of such time to ameliorate the stress that stems from the challenging complexity of day-to-day life.

布什维克海滨公园，纽约，美国
Bushwick Inlet Park in NewYork, USA

迁，遗留下废弃破败的厂房。过去人们为了再现昔日的荣耀，会在空地举行盛大的宗教或皇权庆典，修建壮观的纪念碑，但这些传统的方式已与现代文化模式格格不入。

市中心甚至市郊建筑功能的萎缩不仅仅造成空间闲置，还使整个地区变得破败，这些地方本身一文不值，还因污染和潜在的危险使周围环境恶化。这些地区已经严重退化，无法恢复原有功能，但是如果我们跳出常规的传统规划定式，会发现它们仍有地理位置的优势，具备成为城市结构新起点的潜力，这将以公众意愿为基础，自发性社会力量聚集在一起，可以影响上述地区的发展，产生不同的城市格局。

在经济方面对个人和集体的预期将越来越复杂，变化越来越大，使20世纪理性主义文化所推崇的区域规划理论与实践变得不合时宜，这种呆板的地区规划类型过于庞杂，无法演绎飞速发展的变革。在这场变革中，市民的动态活动改变了生活设施的常规定义。区域规划的滞后显而易见。

促进新公共功能的开发，让废旧场所内的活动焕发生机，都需要想象力。因此，建立新型的城市生活设施是一项艰难的挑战。把原来的工业区改造成专门的社会活动场所是一次有趣的试验，它可以将城市网络在新的基础上重新连接，体会已有空间和闲置空间，已有结构和新建结构以及功能和环境的关系。

在将新建场所纳入城市的过程中，在建成环境内部寻找自然因素的怀旧情结至关重要，因为这种怀旧情结和理想化的自然观不得不在狭小的空间和强烈的人工建筑特征面前妥协，观察这些情结和自然观念如何在城市的实景实物中实现是非常有意思的。

毫不例外，把城市肌理进行整合的主旨是重新审视现有部分。城市

At the same time, the transition from an industrial to a post-industrial society has led to a relocation of production sites that, along with the simultaneous expansion of the metropolis, has left behind places devoid of function and even, often, of shape. Cultural models of modernity also exclude the recovery of such sites through the traditional custom of establishing an astonishing monumentality using the open spaces associated with rhetorical celebrations of religion or tyranny.

This withdrawal of structural functions from downtowns and even from urban suburbs has created not just simple voids, but highly degraded sites that, besides having become worthless in themselves, degrade their context via their polluting quality and their potential for danger. Degraded sites no longer recoverable to their original functions but having the topological potential to become starting points for urban configurations not easily predictable under orderly and deterministic traditional planning yield topologies with variable geometry, shaped by spontaneous social forces around a nuclei of aggregation based on public initiative.

The growing complexity and rapid change of individual and collective economic expectations render inadequate the theory and practice of zoning, so dear to the rationalistic culture of the twentieth century. The sort of planning which worked for homogeneous, rigid areas is simply too cumbersome to interpret a rapid evolution in which the very meaning of the term amenity is routinely modified by citizens, the real authors of all this dynamic transformation. Zoning's decline is thus now evident.

For all these reasons, the creation of new forms of urban amenities has become a formidable challenge in terms of imaginative effort devoted to facilitating new public functions and reclaiming the unpleasant legacy of previous activities. The planning and conversion of brownfields as places for specialized social activities is one of the most interesting laboratories for reconnecting the urban network on a new basis and experiencing new relations among full and empty spaces as well as preexisting and new structures, functions and contexts.

In the process of including these new places in the city, the not necessarily nostalgic desire to find naturalized elements, even inside the built environment, plays an important role. It is interesting to observe how this desire and the concept of nature materializes in urban places and objects, as they come to terms with confined

街道穹顶，哈泽斯莱乌，丹麦
Streetdome in Haderslev, Denmark

可持续性发展和建立相关自治建筑的理念在现实中并不可行。现行举措倾向于重新思考社会在一般模式下相对保守的解决方式，先前的实例表明，以往模式的危机并没有产生新的、行之有效的理念。

很明显，之所以这样保守，原因之一是决心限制资源的消耗，包括土地资源、能源和人力，尽管这些考虑合情合理，但这种曾经推动人类进步的乌托邦情绪终将谢幕。如今，有些建筑名人的大作曲解了这样的情绪。新建筑不是现代法老遗址，不是要含蓄地展示奢华与不同寻常，而是要展示能体现和鼓励新功能和新关系的新型理念。简单来说，我们认为本文项目所代表的趋势是一种反思，可以产生和加强新的城市理念，这是一种战术措施，而不是战略措施。

布鲁克林滨水区废弃场地的改造工程便是其中典范。Kiss+Cathcart把曾经的大型工业停车场改造成布什维克海滨公园，这是纽约东河滨水区大型改造的一期工程。

曾经云集在周边大片地区的工厂和商店渐渐搬迁到更理想的地方，只剩下锈迹斑斑的围栏和破败的停机坪，场地一片萧条。精美并富有创意的涂鸦试图给这片衰败的土地带来生机。从这些涂鸦可以看出，某些民间人士来到这片区域，发现它潜在的价值。

项目整合了两处截然不同的区域。水边和曼哈顿高楼大厦对面的绿地没有与曼哈顿的天际线和河流产生冲突，而是变成一片空地。从河边望去，空地像一个自然的缓坡，上面有醒目的几何形状的步道。社区中心嵌在与操场相连的草坡下面，周围是建筑。缓坡上巨大的圆坑变成露天广场，广场与社区中心的屋顶表面相连。

保留下来的巨大仓库分布在公园两侧，使布局井然有序，从而找回了昔日荣耀。

spaces and the characteristics of strong artificial structuring. With few exceptions, however, the central theme of this process of re-stitching the urban fabric has been a strong reconsideration of the existing. There has been practically no room for conceptions of a continuously evolving city and the creation of related autonomous architectures. What prevails seems to be a conservative answer, already seen in other historical eras, to a moment of rethinking the general model of society, in which the crisis of the previous model has not yet resulted in a total, positive new idea of itself.

It is clear that among the reasons for such stodginess has been a strong resolve to limit resource consumption, whether of soil, energy or personal time, but despite the undeniable reasonableness of such concerns, the result seems to have been the sunset of certain utopian impulses that have always underlain great advancements of humankind. Such impulses are misinterpreted, today, in the spectacular examples of the establishment of archistars. A new architecture does not necessitate a secluded spectacularity of extravagant and exceptional items, as of modern pharaonic monuments, but the conception of new forms that embody and encourage new functions and relationships. In a nutshell, we can define the trend represented by the projects examined here as a moment of reflection awaiting the birth and consolidation of a stronger new idea of the city, a tactical approach more than a strategical one.

Exemplary in this regard is the requalification of a discarded area on the Brooklyn waterfront. The project of Kiss + Cathcart transforms what was a vast industrial parking lot into the Bushwick Inlet Park, a pioneering intervention of a larger redevelopment of New York City's East River waterfront.

The extended surrounding area presents itself as a depressing sequence of abandoned sites, once home to commercial and industrial activities, whose gradual removal to more suitable locales has gradually left behind a landscape of rusty fences and degraded aprons. However, it is clear that informal forces wind through the area that find in it potential value, as is vividly shown in the elegant and creative graffiti that attempt to restore a soul to the degraded environment.

The project is articulated into two quite distinct parts. A green playground facing the waterfront and the skyline of Manhattan refrains from confronting the Manhattan skyline and the river, in-

马戏艺术学院，欧什，法国
Circus Arts Conservatory in Auch, France

CEBRA+Glifberg+Lykke在丹麦哈泽斯莱乌设计的嵌入结构更注重为街头运动或无组织运动创造机会，提出与非正式运动相协调的创作方案。和布什维克海滨公园一样，不同建筑风格的物体和因素参照自然形式组合在一起。

对住宅区和服务区进行大规模改造后，哈泽斯莱乌的海滨修建了"街道穹顶"，它是一处文化中心，为居民提供了各种便利设施。重新装饰的旧筒仓作为新地标被保留下来。街道穹顶的地面呈凹形，适用于多种类型的运动。凹形地面与主要起保护作用（而非一座建筑）的开放式圆顶相呼应，圆顶下人们还可以在人造墙上练习攀岩。

Tod Williams Billie Tsien建筑师事务所设计的美国纽约布鲁克林区前景公园湖滑冰场采取了比较传统的方式。事实上，该建筑不仅为定期滑冰训练和曲棍球训练提供场所，也成为前景公园湖两岸大型公园的附加设施。公园的特色是很容易让人联想到大自然。

公园的冰上运动区进行了深度改造，该项目不只是一处封闭的空间，其功能更加复杂，能全年提供各种运动项目。和上个例子一样，该建筑是开放式的，与室外空间相连接。顶棚嵌板的图案好像冰鞋滑过冰面留下的痕迹，让人联想起夜空，这种设计使顶棚的嵌板显得不那么突兀。粗糙的花岗岩石墙、石柱与柔软的公园植被形成强烈对比，协调统一。

在法国纳什镇马戏艺术学院，Doazan + Hirschberger & Associés重新发现了马戏的真谛——马戏不仅仅是偶尔的杂耍和杂技，还为娱乐和邻里聚会提供了永久的机会。这表明，为城市复兴、社会和谐设想新颖的、不同寻常的特色是完全有可能的。

固定的木质马戏大篷是醒目的地标，旁边是不显眼的Espagne兵营，

stead limiting itself to an unbuilt space, mimicking a natural slope from the point of view of the river, although marked by pathways as strong geometric signs. The community center is wedged under the grass carpet in continuity with the playground and dimensionally confronts the built surroundings. Ample rounded holes in the slope generate open plazas that connect the top surfaces with the community center.

One result is that the surviving huge warehouse flanking the park at last regains its lost dignity as it assumes a role of bringing order to the context.

The intervention of CEBRA + Glifberg + Lykke in Haderslev, Denmark, is more focused on creating opportunities for street or otherwise unorganized sports and proposes a compositional scheme in tune with the informal nature of the activities it addresses. Like Bushwick Inlet Park, it is a mix of architecturally defined objects and elements that in some way refer to natural forms.

The Streetdome sits in an area along the Haderslev Fjord, affected by an intense remodeling which includes residential complexes and services, with respect to which it is proposed as a hub for culture and amenities. The project retains the landmark of a newly decorated old silo. It emerges as a carved surface suitable for many sporting activities, in continuity with the open dome which is, in this way, more a protection than a real building. Under the dome it is possible to practice climbing on an artificial wall.

A more traditional approach was adopted by Tod Williams Billie Tsien Architects for the ice skating rink at LeFrak Center in Lakeside, Brooklyn, NY. In fact, besides being designed for the well-established practice of ice skating and hockey, the complex is an additional amenity for a vast parkland, already characterized by strong connotations of naturalness, along the banks of Prospect Park Lake.

The project is part of a profound reshaping of a park area already dedicated to ice sports, and is enriched with more complex functions able to adapt themselves over the seasons, moving beyond the concept of a closed box. As in the previous case, the building is open and in continuity with the outdoor space. The rigid sign of the ceiling slab is mitigated by a treatment that nods to the traces left by skates on ice but which also recalls a night sky. Granite walls and pillars arise in a dialectic with the soft design of the park's plants.

Allez-Up攀岩中心，蒙特利尔，加拿大
Allez-Up Climbing Center in Montreal, Canada

兵营被重新设计，里面有剧团工作室、演播室和宽敞的排练厅。大篷和兵营之间的公共空地被称作"la strada"，意大利语意为"街道"，显然向Fellini曾经讲述街头艺术家的电影致敬。

马戏团的长期驻扎给小镇带来活力，小镇也对大规模改造项目充满兴趣。在全球化时代，快速的变革甚至在非正式活动和新兴运动中快速地创造出参照模板。如果哈泽斯莱乌的街道穹顶的攀岩墙是众多特色之一，那么Lanz + Mutschlechner + Wolfgang Meraner在意大利南提洛尔Brixen镇的设计和Smith Vigeant建筑事务所在加拿大蒙特利尔的设计不仅为大众消遣活动提供了固定的场所，还通过其清晰的建筑模式来决定它们的形态特征。有趣的是，蒙特利尔"Allez-Up"攀岩区与周围环境的联系如此紧密，以致于清楚地暗示了该地点以前的用途：攀岩墙的白色表面让人联想起先前在仓库里存放过的白糖。

毫无疑问，这些项目都介绍了维持环境可持续性的方法，这些是项目获得广泛共识的必要条件。

很明显，人们想到处理空闲时间的新方法，渴望提升个人和集体生活的动力，要重新定位失去往日光辉的城市区域，这一切都催生了清晰而又复杂的创新机会，在千年之交壮观的建筑风格之外，人们仍然可以从中找到新的灵感和解决方法。

In the Circus Arts Conservatory, in the town of Auch, France, Doazan+Hirschberger & Associés rediscover the circus arts as a permanent opportunity for recreation and community gatherings and not simply as the occasional activity of acrobats and jugglers. Such is a clear example of the possibility of imagining new and unusual features for urban revitalization and for community cohesion.

The stationary wooden framed circus tent, a highly recognizable landmark, sits alongside the neglected Espagne Barrack, redesigned to accommodate resident troupes, studios and a large rehearsal room. The public space between the tent and the barrack is allusively called "la strada", Italian for the street, an evident homage to Fellini's movie dedicated to street artists.

The stable presence of circus troupes represents a way to revitalize this part of the town, interested in a large replanning program. The rate of change of the era of globalization, however, tends to quickly create reference models, even in the context of informal movements and the emergence of new practices. If in the above case of the Haverslev Streetdome the climbing wall was one of many features, two examples in Brixen, South Tyrol, Italy, by Lanz + Mutschlechner + Wolfgang Meraner, and Montreal, Canada, by Smith Vigeant Architectes, focus specifically on this activity which, in addition to consolidating the sites as homes to a popular pastime, determines their morphological features via the emergence of a rather well defined architectural model. Interestingly, in the case of the climbing facility in Montreal, "Allez-Up", the gesture of connection with the context goes so far as to playfully allude to the previous destination of the site: the white faceted climbing walls recall the sugar once contained in the former silos.

It goes without saying that all these projects introduce measures aimed at environmental sustainability, which has become a condition recognized as essential for a project to obtain broad consensus. It is clear that the definition of new ways to dispose of free time, the desire to expand the motivations of individual and collective existence, and the need to rethink parts of the city that have lost their original meaning, are all producing very articulated and complex moments of creativity from which it is still possible to await new intuitions and solutions in contrast to the spectacularity that has characterized architecture at the turn of the Millennium.

Aldo Vanini

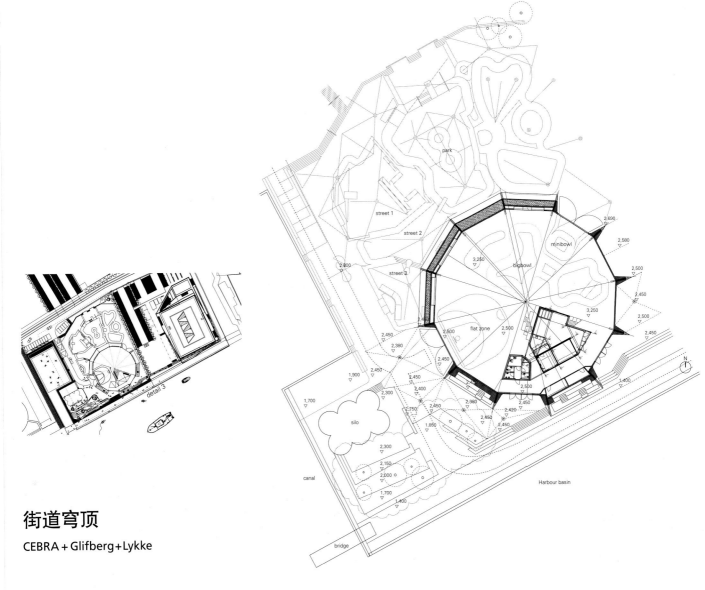

街道穹顶

CEBRA + Glifberg+Lykke

    街道穹顶是一个广阔而又独一无二的城市景观，其中还包括一个4500m²的滑板公园，供街头篮球、跑酷、攀岩、轻艇水球等休闲娱乐活动使用。

    街道穹顶项目的总体目标是建立城市自主体育运动场的新标准。它为不同年龄，不同技术水平，以及不同文化层次的人们提供了一个开放性的娱乐活动场地和社交聚会场所。这个项目大概始于十年前，是滑板城市联合会主席，也是当地积极分子Morten Hansen的创意。他旨在打造一个新型且覆盖各个领域活动和娱乐的文化枢纽。

    建筑师Glifberg+Lykkes设计的滑板公园整体上分为三部分，广场围绕着人流区域以及沟渠和河岸为特色的公园区延伸。随着使用者技能水平和想象的不断发展，公园将不断地挑战运动员的身体能力和创造能力。

    街道穹顶本身基于CEBRA事务所的圆顶大厅概念而设计的。为了减少运营成本，大厅没有供暖设施。穹顶的外形使建筑表面区域最小化，采光也以自然光为主。屋顶跨度大约为42m，形成一个巨大的且没有承重结构的开阔空间。建筑适合举办各种体育活动，其中以开展滑板、攀岩和街头篮球为主。

    圆顶的设计非常适合场地自身的条件，也适用于展现滑板和街头文化的特殊功能。从外部看，圆顶是滑板公园的实际组成部分，就像从混凝土景观中长出的蘑菇。该设计成为公园内的一个实用部分，人们可以在建筑边缘的河岸、楼梯和边缘斜坡上轮滑。从内部看，地面上有精心挖造的水池，旁边是街头篮球场以及一个位于中心的巨型大圆石，圆石设有表演平台、坐椅和洗手间。宽阔的大门朝外打开，将滑板公园和建筑内部连接为一体，形成了整个公园的无缝式流动。街道穹顶形成了一种连续且可变的空间轨迹，赋予斯堪迪纳维亚一个特殊而又蕴含着无数可能的现代街道运动竞技场。

    圆顶结构的胶合木拱顶位于混凝土塔式结构之上，其跨度为42m，高度为10m。

    屋顶/立面由预制的木构盒式结构建成，木构盒子由两个填充屋顶的油毛毡和粗糙的表面进行密封，粗糙的表面生长着苔藓景天科植物。街道穹顶的屋顶覆盖着种好的苔藓景天科植物，包括各种景天草（苔景、玉米石、高加索景天植物等）和同时附生的苔藓。

    绿色的屋顶同时在美学上和功能方面发挥作用。从视觉方面看，屋顶会随着季节和天气改变颜色。植物的外观也会随着当地气候和施肥的影响而改变。

    从功能上来说，绿色的屋顶提供了更适宜的城市气候以及室内气候。因为大厅并没有设置保温设施，且仅仅使用天然的通风设施，所以绿色植物能够避免夏天室内过热，起到很好的隔声、过滤空气以及减少颗粒污染的作用。除此之外，屋顶还能够吸收雨水，从而避免雨水从屋顶流下而影响到滑板公园。

Streetdome

Streetdome is a vast and unique urban landscape for activity and recreation including a 4,500m² skate park, and facilities street basket, parkour, boulder climbing, canoe polo etc.

Its overall ambition is to set new standards for urban arenas for unorganized sports. Streetdome is an open playground and social

meeting place for different ages, skill levels and cultures. The project started out almost ten years ago as an initiative by local activist and chairman of the Skate City Association Morten Hansen, whose vision was to create a new cultural hub that incorporates a vast area for activity and recreation.

Glifberg + Lykke's design for the skate park is divided into three overall sections consisting of the street plaza, which extends around the transitions-based flow section and the park section characterized by ditches and banks. The park develops over time along with the users' skills and imagination and continue to challenge the athletes on sporting as well as creative levels without an expiration date.

The Streetdome itself is based on CEBRA's igloo hall concept. To reduce running costs, the hall is unheated and lit primarily through

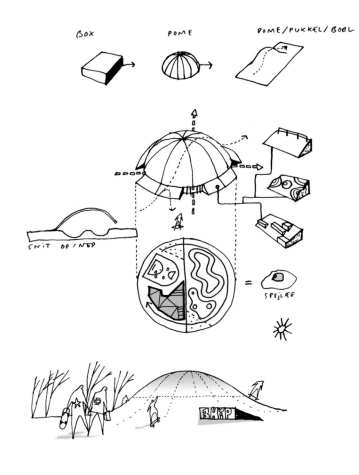

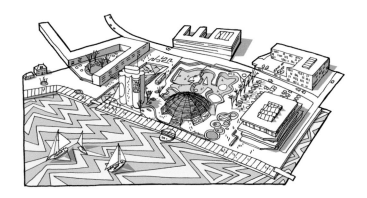

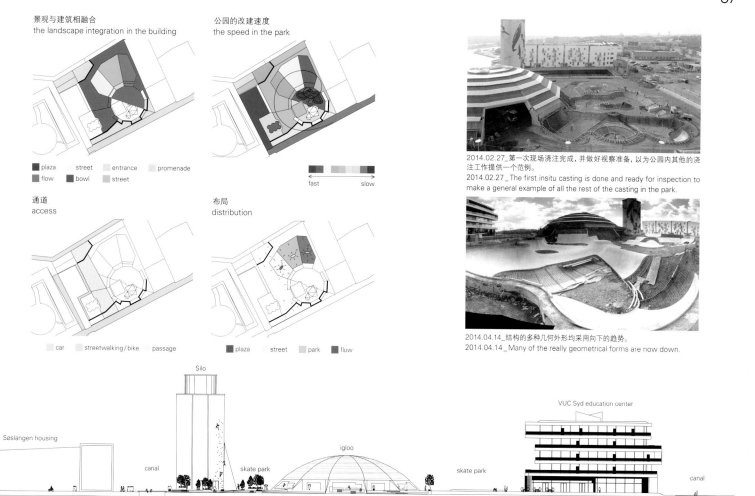

2014.02.27_第一次现场浇注完成,并做好视察准备,以为公园内其他的浇注工作提供一个范例。
2014.02.27_The first insitu casting is done and ready for inspection to make a general example of all the rest of the casting in the park.

2014.04.14_结构的多种几何外形均采用向下的趋势。
2014.04.14_Many of the really geometrical forms are now down.

东南立面 south-east elevation

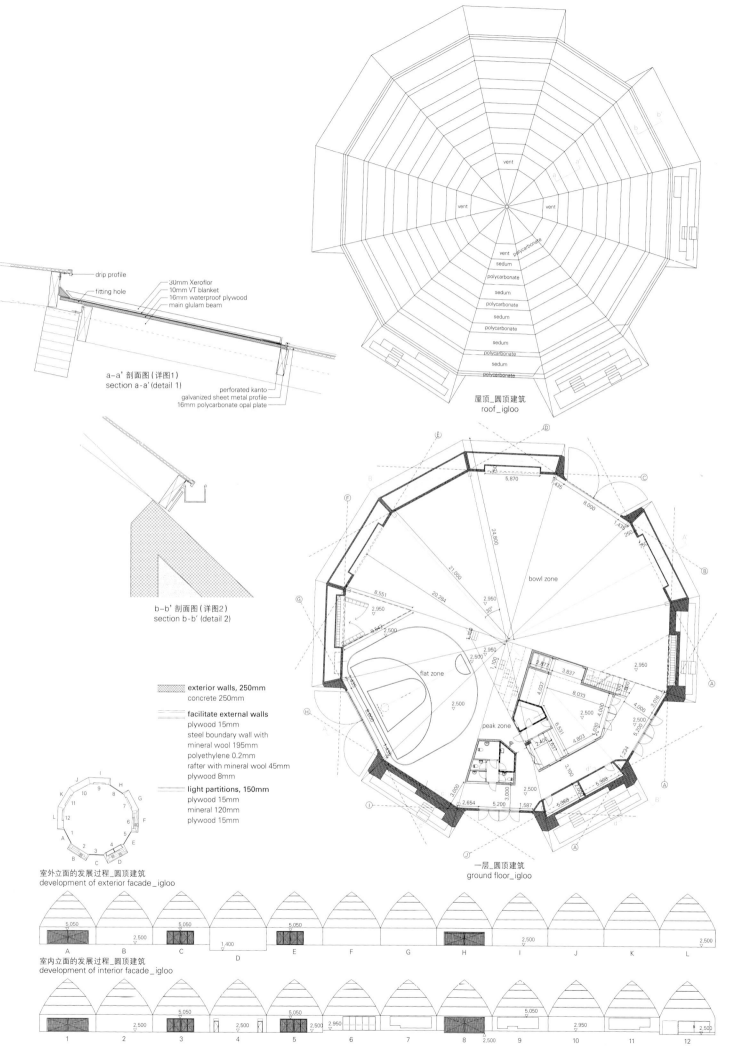

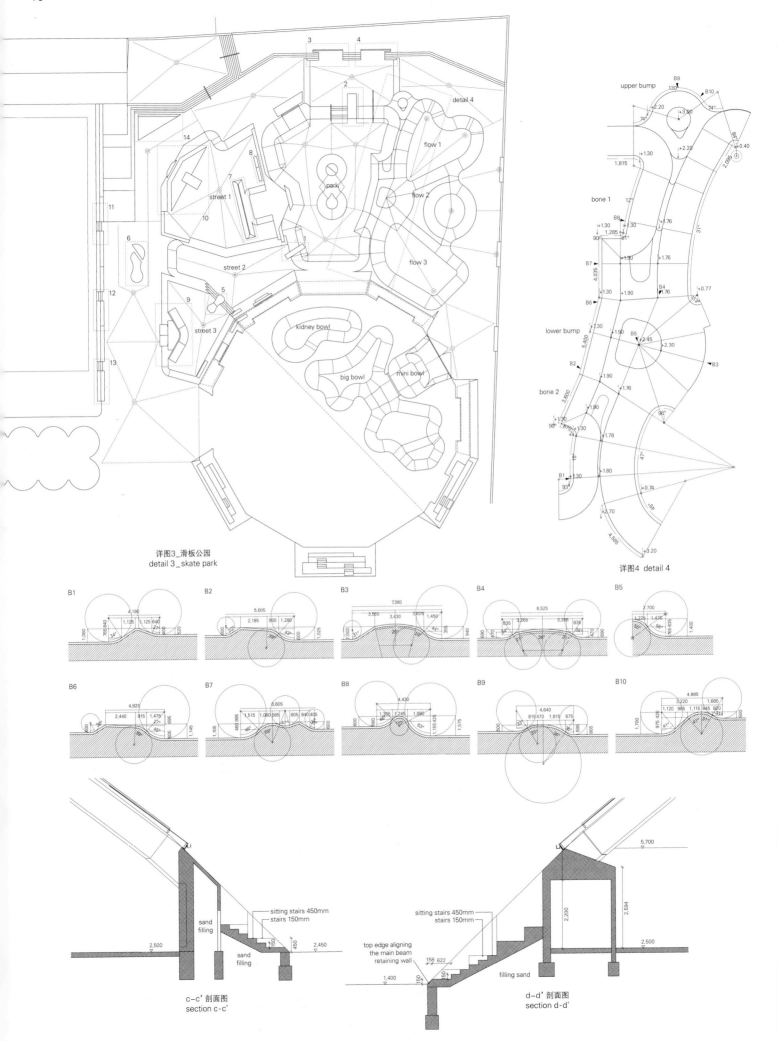

daylight while the building's surface area has been minimized through its dome shape. The roof spans around 42m and allows for a large open space free of load-bearing structures, which can be fitted out for any kind of sports and activities, in this case of skateboarding, boulder climbing and street basket.

The Igloo is adapted to both the site and the specific function of skateboarding and street culture. On the outside, the dome is an actual part of the skate park, growing out of the concrete landscape like a mushroom. The structure is designed as a functional part of the park to skate on with banks, stairs and slopes along the rim. Inside, a series of pools are scooped out of the floor next to a street basket court and a central boulder structure containing a performance platform, seating and bathrooms. Wide gates open to the outside connecting the surrounding skate park with the inside floor, creating a seamless flow through the entire park. Streetdome forms one continuous and varied spatial course, which gives Scandinavia a modern street sport arena with unique features and possibilities.

The dome structure's laminated wood arches, which are mounted on concrete pylons, span 42m and rise to a height of 10m.

The roof/facade is constructed of pre-fabricated timber cassettes, which are sealed with 2 x roofing felt and rough surfaces for the moss-sedum elements. The Street Dome's roof is covered with pre-fabricated moss-sedum elements, which consist of a variety of stonecrops(Sedum acre, Sedum album, Sedum spurium etc) and spontaneously occurring moss.

The green roof serves both aesthetic and functional purposes. Visually, the roof will change color according to season and weather conditions and the vegetation's appearance will develop over time influenced by the local climate and fertilization.

Functionally, the green roof contributes to a better urban as well as indoor climate. As the hall is uninsulated and only uses natural ventilation the vegetation prevents overheating during the summer, acts as sound insulation, and filters the air and reduces particle pollution. In addition, the roof is able to absorb rainwater, which then evaporates instead of running off the building and onto the skate park.

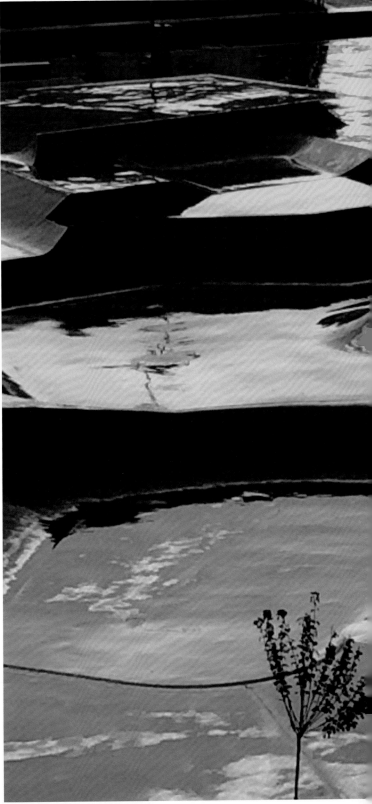

照片提供：© Jacobesben (courtesy of the architect)

项目名称：Streetdome
地点：Haderslev, Denmark
建筑师：CEBRA + Glifberg+Lykke
设计团队：
CEBRA_Mikkel Frost, Lars Gylling, Glifberg-Lykke_Ebbe Lykke, Rune Glifberg
Interior, 项目开发：Morten Hansen
工程师：Rambøll / 施工：Hoffman, Grindline
出资人：A.P. Møller Fonden
用地面积：4,500m² / 总建筑面积：1,500m²
材料：indoor_concrete, laminated wood trusses, plywood, exterior_concrete (shotcrete), moss-sedum, polycarbonate panels
设计时间：2011 / 竣工时间：2014
摄影师：
Courtesy of the architect-p.35, p.36~37, p.41 middle, bottom, p.44, p.45
©Arto Saari (courtesy of the architect)-p.38, p.41 top

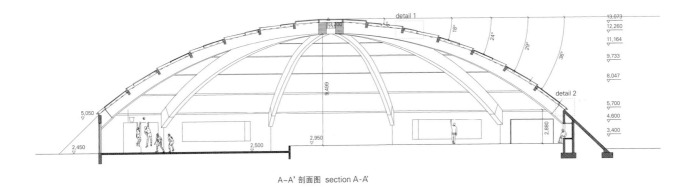

A-A' 剖面图 section A-A'

B-B'剖面图 section B-B'

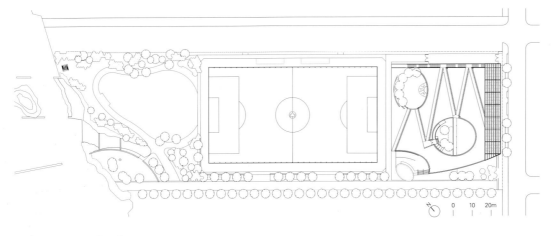

布什维克海滨公园

Kiss+Cathcart

　　布什维克海滨公园将布鲁克林滨水区由一个废弃的工业地带变为一座公共公园。它坐落于布鲁克林的威廉斯堡地区，其设计是河岸东部滨水区域重建计划的第一步。设计团队精心将一系列的运动场地、社区设施以及纽约中央公园的维护设施和操作设备融入到这处面积为2.5公顷的公园内。

　　公园在西侧将建筑围起来，将建筑变成一座绿丘，使场地对公众全面开放。弯曲的小径为体育运动委员会的成员们提供了通往顶部的通道，一个大型木质天篷则提供了遮阴，人们在此可从球场一路看向曼哈顿的天际线。公众和员工还可以通过下方的街道一侧直接进入建筑内。

　　占地面积为1236m²的公园分为南北两个区域。北区是为公园的行政区和休闲区服务的维护设施，南区是由非盈利的空地联盟所运营的社区活动中心，每处功能区都拥有独立的入口。

　　该项目最初的规划要求是希望在公园里建造LEED银奖水准的建筑。然而，设计团队却强烈地感觉到位于敏感的滨水地带的公共区要集用途、规划以及环境于一体，同时旨在达成更高的环境标准。公园在建筑上方徐徐升起，使公共空间的面积最大化。在面向西北的绿色屋顶下，建筑受到了大地的庇护，提供了遮阴，因此东南和西南侧只能接收到有限的日光暴晒。建筑内设有抽象的、可循环使用的、像树根一样的映射在大厅天花板上的铝质吊灯，且所有的可用空间，包括大堂，都有充足的自然光。

　　所有的电气设施，主要包括太阳能动力机械系统，都使用了不占用屋顶空间的冷水机组、烟道和通风孔，并且该系统无噪音，零排放。屋顶花园是利用收集的雨水和游乐园中喷洒的废水来灌溉的。

　　该建筑被设计为一处功能性强且美观的公共空间。室内与室外都与城市和公园的不同条件相呼应。在公园一侧，建筑与景观融为一体，而在街道一侧，立式木质遮阳柱廊赋予了公共立面特色，标志着布鲁克林北部最重要的新建公园的入口所在地。

　　布什维克海滨公园南部边缘的北九大道作为一条延伸的人行道，突出了现有街道的布局，并将该布局延伸至海中。公园旨在从南侧延续州立公园的道路，并在北部与总体规划的后期项目相连接。

　　绝大多数游客乘坐公交车或是骑自行车前来。这里的街区较少，设有公交车、地铁、河岸东部的轮渡码头、自行车道、公共自行车共享码头，此外还有许多人居住在这里。仅有的停车场可供维护人员使用，一共有八个区域，六个在北院的屋顶花园下方，两个在车库。

Bushwick Inlet Park

Bushwick Inlet Park transforms the Brooklyn waterfront from a brownfield industrial strip into a public park. Located in the Williamsburg section of Brooklyn, the design is the first step of a wa-

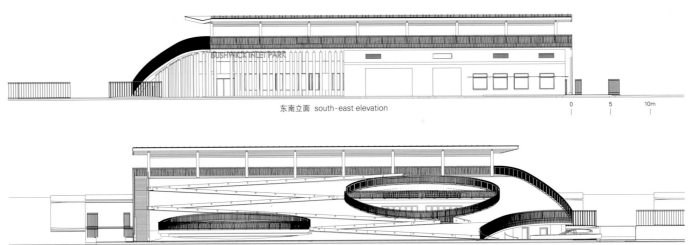

东南立面 south-east elevation

西北立面 north-west elevation

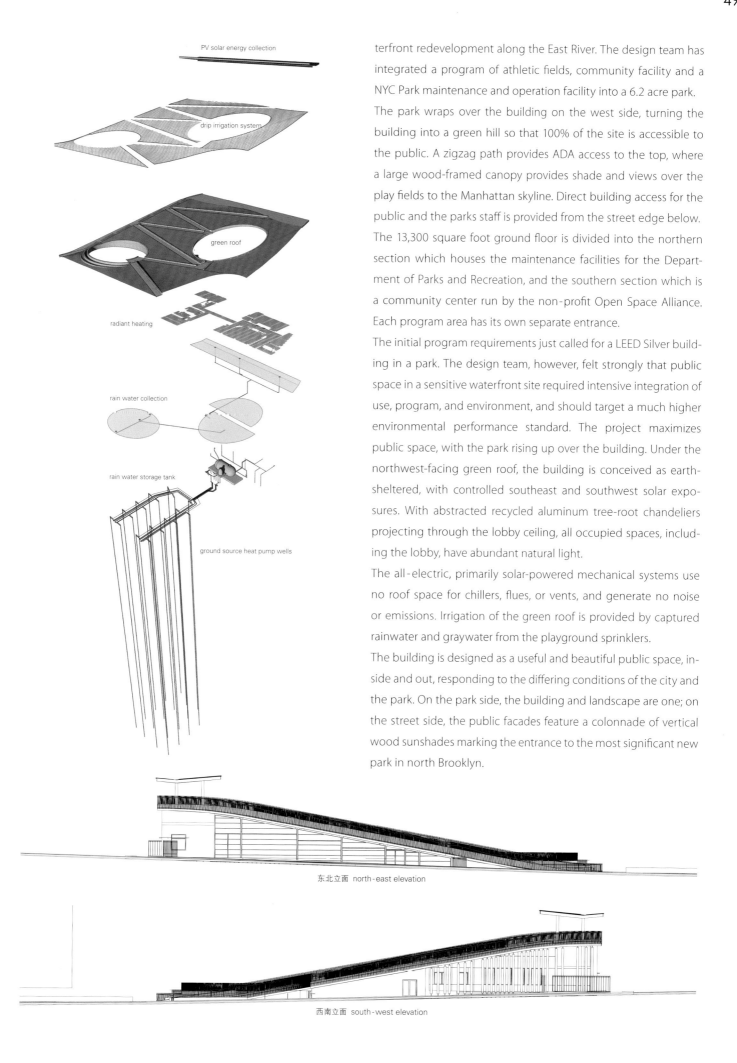

terfront redevelopment along the East River. The design team has integrated a program of athletic fields, community facility and a NYC Park maintenance and operation facility into a 6.2 acre park. The park wraps over the building on the west side, turning the building into a green hill so that 100% of the site is accessible to the public. A zigzag path provides ADA access to the top, where a large wood-framed canopy provides shade and views over the play fields to the Manhattan skyline. Direct building access for the public and the parks staff is provided from the street edge below. The 13,300 square foot ground floor is divided into the northern section which houses the maintenance facilities for the Department of Parks and Recreation, and the southern section which is a community center run by the non-profit Open Space Alliance. Each program area has its own separate entrance.

The initial program requirements just called for a LEED Silver building in a park. The design team, however, felt strongly that public space in a sensitive waterfront site required intensive integration of use, program, and environment, and should target a much higher environmental performance standard. The project maximizes public space, with the park rising up over the building. Under the northwest-facing green roof, the building is conceived as earth-sheltered, with controlled southeast and southwest solar exposures. With abstracted recycled aluminum tree-root chandeliers projecting through the lobby ceiling, all occupied spaces, including the lobby, have abundant natural light.

The all-electric, primarily solar-powered mechanical systems use no roof space for chillers, flues, or vents, and generate no noise or emissions. Irrigation of the green roof is provided by captured rainwater and graywater from the playground sprinklers.

The building is designed as a useful and beautiful public space, inside and out, responding to the differing conditions of the city and the park. On the park side, the building and landscape are one; on the street side, the public facades feature a colonnade of vertical wood sunshades marking the entrance to the most significant new park in north Brooklyn.

东北立面 north-east elevation

西南立面 south-west elevation

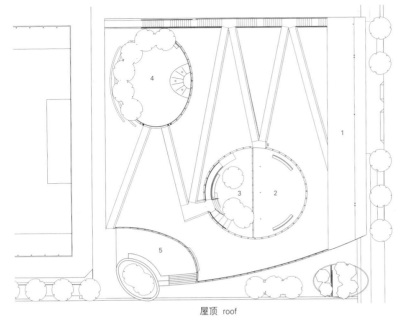

屋顶 roof

1 上层广场	1. upper plaza
2 日光浴甲板	2. sundeck
3 多功能广场	3. multipurpose plaza
4 游乐场	4. playground
5 舒适的车站广场	5. comfort station plaza

一层 first floor

1 公共入口	1. public entry
2 大堂	2. lobby
3 教室	3. classrooms
4 多功能室	4. multipurpose rooms
5 社区办公室	5. community office
6 公共卫生间	6. public WC
7 午餐室	7. lunch
8 机械间	8. mechanical
9 车库	9. garage
10 衣帽间	10. lockers
11 维修&操作室入口	11. M&O entry
12 办公室	12. offices
13 覆顶的公园	13. covered parking
14 存储室	14. storage
15 公园卫生间	15. park WC

North 9th Street was extended as a pedestrian path at the southern edge of Bushwick Inlet Park, reinforcing the existing street grid and continuing it to the water. The park was designed to continue the State Park's pathways from the south and to link to the future phases of the Master Plan to the north.

The vast majority of visitors come by public transport or bicycle. Within a few blocks there is bus, subway, an East River Ferry dock, a bike path, a Citi Bike share dock, and many residents. The only parking provided is for park maintenance personnel, with six spaces under the green roof in the north yard, and two inside in the garage.

项目名称：Bushwick Inlet Park / 地点：86 Kent Avenue, Brooklyn, NY 11211, United States
建筑师：Kiss + Cathcart
设计合作者：Gregory Kiss / 项目建筑师：Clare Miflin
项目经理：Randee Stewart / 建筑师：Heather McKinstry
施工团队：URS Corporation
结构工程师：Robert Silman Associates
土木工程师：Langan Engineering & Environmental Services Inc.
MEP工程师：A.G. Consulting Engineering, PC
照明设计师：AWA Lighting Designers 景观屋顶设计顾问：Roofmeadow
场地工程师：Wesler-Cohen Associates
LEED：design_Community Environmental Center Inc.; construction_Taitem Engineering, PC
能源建模委托机构：Taitem Engineering, PC
雨水收集设计师：Geosyntec Consultants
估价：Accu-Cost Construction Consultants, Inc.
监理：Construction Specifications, Inc.
景观建筑师：Starr Whitehouse Landscape Architects, Planners PLLC
业主：New York City Department of Parks and Recreation
用地面积：25,000m² / 总建筑面积：1,400m²
设计时间：2008—2009 / 施工时间：2009 / 竣工时间：2014
摄影师：©Paul Warchol (courtesy of the architect) (except as noted)

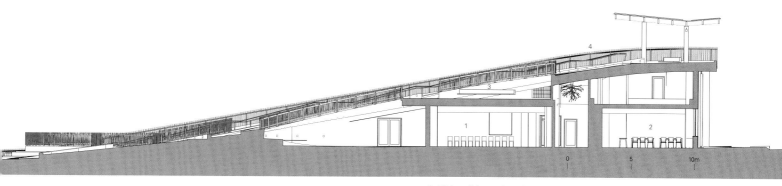

1 多功能室 2 教室 3 日光浴甲板 4 上层广场
1. multipurpose rooms 2. classrooms 3. sundeck 4. upper plaza
A-A' 剖面图 section A-A'

湖滨的勒弗拉克中心
Tod Williams Billie Tsien Architects

前景公园是一座位于纽约市布鲁克林区的面积为2 367 411m² 的公共公园。它是弗雷德里克·劳·奥姆斯特德和卡尔弗特·沃克斯在完成曼哈顿中央公园的设计之后的又一力作。1961年,为了扩建出一个溜冰场,拆除了音乐岛跟滨海大道。当年的马车广场也成为一个大型停车场。前景公园联盟和纽约市政娱乐与公园管理委员会共同资助改造了105 218m² 的公园绿地,建造了勒弗拉克中心,使这片区域焕然一新。原有的溜冰场被拆除,在原来的地方又重建了音乐岛和滨海大道,勒弗拉克中心也取代了那个大型停车场。这里是一个终年开放的娱乐和休闲胜地,也可用作举办活动的场地。

勒弗拉克中心看起来是由两个单层花岗岩覆盖的体块组成的,呈L形。不规则排列的柱子支撑了一个33m宽、63m长的天篷。该建筑部分由土覆盖,由此引导游客去参观屋顶露台。这些屋顶露台与公园的小径相通,游客通常不知不觉中便走到了天台,而没有意识到此时建筑已在下面。

曲棍球场被建筑体块围合,位于一个天篷之下,其规格与普通的球场无异,在温暖的天气里还可以用作轮滑场地。紧挨着曲棍球场的是一个椭圆形娱乐场,在夏天,这里便成为了一处喷泉游乐场所。建筑内设有售票区、滑冰器具租赁中心、办公室、咖啡厅、宴会室、卫生间和机械室。

沿着曲形墙壁行走,游客们便进入了勒弗拉克中心。建筑的墙面覆盖了9cm厚的石板,石板是由魁北克的劳伦系绿花岗岩切割而成的。勒弗拉克中心并没有正式的入口。游客既可以在回收的柚木建成的青铜色桥下穿过,也可以在入口租用轮滑器具。坚固的石墙设有洞口,打破了整体的连续性,人们在洞口处可以看到周围的景观。27.4m长、2.4m宽的装饰性瓷砖墙面镶嵌在入口的墙壁上,其选取的颜色和设计的图案使人们感受到了四季的更替。

北面的体块设有租赁与更换区,这里还为滑冰的人们或普通游客提供了更衣室和卫生间。勒弗拉克中心有望获得LEED金牌认证,它将成为世界上唯一一个获得LEED认证的滑雪场。滑雪中心的制冰系统采用氨,这是一种非常有效的制冷剂,而且不会危害臭氧层。

天篷由十个不规则排列的柱子支撑。下面为嵌有耀眼金属凹壁的合成灰泥墙。该墙体涂成午夜蓝色,其表面的雕刻设计灵感来自于人们滑冰的步法模式。屋顶上种有景天属植物。

屋顶露台的视野十分宽阔,人们在此可以俯瞰溜冰场,还可以看到远处的公园与湖泊。天气暖和的时候,北露台开放,而东露台的屋顶上放置了临时的亭子、桌子、遮阳伞和椅子,形成了一座屋顶啤酒花园。一座桥将这两个露台连接起来。

LeFrak Center at Lakeside

Prospect Park is a 585 acre public park in the New York City borough of Brooklyn. It was designed by Frederick Law Olmsted and Calvert Vaux after they completed Manhattan's Central Park. The addition of a new skating rink in 1961 entailed the bulldozing of Music Island and demolition of the Esplanade. The carriage concourse became a large parking lot. The Prospect Park Alliance and NYC Department of Parks and Recreation collaborated to finance twenty-six acres of parkland renovation to original vision and the creation of the LeFrak Center. The old rink was demolished, Music Island and the Esplanade were reconstructed in its place, and the LeFrak Center replaced the large parking lot. The new building is a year-round destination for recreation, relaxation, and events.

The LeFrak Center consists of two one-story granite clad building blocks configured in an L-shape. Irregularly placed columns support an 108-foot wide and 208-foot long canopy. The building blocks are partially covered by earth to allow visitors to walk up to the roofs which act as terraces. These upper level terraces connect with the park pathways so visitors will find themselves at this elevated position without realizing that there is a building beneath. Framed by the building blocks and sheltered by the canopy is a

1 曲棍球场 2 前景花园湖泊 3 停车场
1. hockey rink 2. Prospect park lake 3. parking
原有的区域（1961） existing (1961)

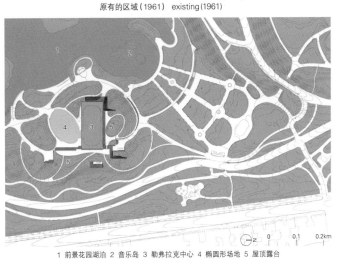

1 前景花园湖泊 2 音乐岛 3 勒弗拉克中心 4 椭圆形场地 5 屋顶露台
1. Prospect park lake 2. music island 3. LeFrak center 4. elliptical rink 5. roof terrace
现在的区域（2013） current (2013)

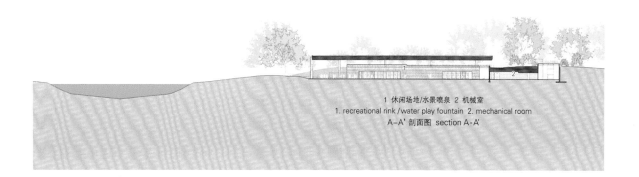

1 休闲场地/水景喷泉 2 机械室
1. recreational rink / water play fountain 2. mechanical room
A-A' 剖面图 section A-A'

天花板_曲棍球场/轮滑场 ceiling_hockey rink/roller skating

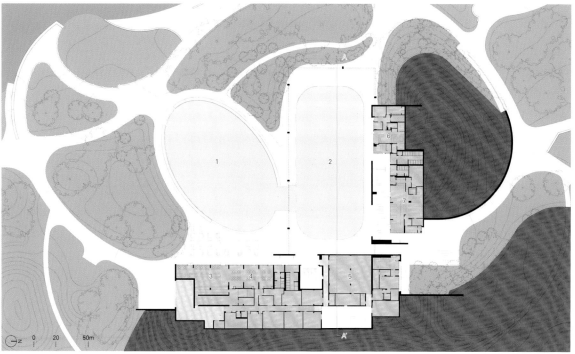

1 休闲场地/水景喷泉 2 曲棍球场/轮滑场 3 咖啡室 4 聚会室 5 机械室 6 办公室 7 滑冰器具租用/更换区
1. recreational rink / water play fountain 2. hockey rink/roller skating 3. cafe 4. party room 5. mechanical room 6. office 7. skate rental/change
一层 first floor

regularized hockey rink which converts to roller skating in warm weather. This rink connects to an elliptically-shaped recreational rink that is a water play fountain in the summer. The building blocks house the ticketing area, skate rental, offices, a cafe, party rooms, rest-rooms and mechanical spaces.

Visitors follow a curved wall to enter the LeFrak Center. Laurentian Green Granite from Quebec cut into three-and-a-half inch thick split-faced blocks clads the building. The building has no formal entry. Visitors can either pass under a bronze and reclaimed teak bridge or purchase a ticket for skate rentals at the building entry. The solid stone walls are interrupted by openings which frame the surrounding landscape. A ninety-foot long and eight-foot tall decorative tile mural is inset in the entry wall. The colors chosen and pattern developed evoke the changing seasons.

In the north block is the skate rental and change area as well as lockers and rest rooms for skaters or general park visitors. The east block holds the mechanical space. The LeFrak Center is on track to receive LEED Gold Certification, upon which it will be one of the only LEED certified skating facilities in the world. The system for ice making utilizes ammonia which is a very efficient refrigerant and does not contribute to ozone depletion.

The canopy is supported by ten irregularly placed columns. The underside is a synthetic stucco accentuated by metallic alcoves. It is painted midnight blue and carved with a design inspired by figure skating footwork patterns. Sedum is planted on the roof of the canopy.

The roof terraces overlook the rinks and allow for generous views of the park and lake beyond. The north block terrace was opened while a temporary pavilion, tables, umbrellas, and chairs are brought out in warmer months to create a roof-top beer garden on the east block terrace. A bridge connects the two terraces.

项目名称：LeFrak Center at Lakeside / 地点：Prospect Park, Brooklyn, New York, USA
建筑师：Tod Williams Billie Tsien Architects
首席建筑师：Tod Williams, Billie Tsien / 项目经理：Andy Kim / 项目建筑师：Elisa Testa
项目团队：Erin Putalik, Nate Petty, Shengning Zhang
施工经理：Sciame Construction / 土木工程师：Stantec / MEP：ICOR Associates
溜冰场内的制冷设计：Van Boerum & Frank Associates / 结构：Robert Silman Associates
视听设计：Acoustic Dimensions / 编码：William Vitacco Associates
委托方：EME Group / 土工设计：Richard Kessler / 制图：Poulin + Morris
景观屋顶设计：Roofmeadow / 厨房设计：Ricca Newmark
景观设计：Prospect Park Alliance Dept of Design and Construction
LEED顾问：7Group / 照明设计：Renfro Design Group
安全顾问：Ducibella Venter & Santore / 监测：Construction Specifications
用地面积：75,000m² / 总建筑面积：34,000m²
设计时间：2006.1—2011.1 / 施工时间：2011.2—2013.12 / 竣工时间：2013.12
摄影师：©Michael Moran/OTTO

欧什马戏艺术学院

Doazan + Hirschberger & Associés

位于西班牙叛逃营旧址的马戏创新研究中心的建立是欧什城市重新规划的第一步。其功能包括常驻剧团使用的房间、工作室、一间大型排练室以及一个可容纳700人的永久性马戏团帐篷。该项目共打造了三个不同的"世界",但是从整体来看,这三个"世界"形成了一处有趣、富有凝聚力的空间。

首先,马戏创新研究中心的各个空间围绕天井而设。天井为半庭院式,是一处居民生活中心,人们在这里可以自由地工作、聊天、追求梦想。

工作室围绕着排练室设置,成为马戏团艺术工作者的中心工作区域。考虑到马戏团员工的工作强度大,且需要注意力高度集中,因此用于表演的、充当舞台背景的塔形结构是一个极其便利的设施,提供了最高的效率。其立面覆盖了落叶松板,一些改造的马厩从中显露出来。

定居于此的马戏团成为了一个强烈且明显的城市标志:不透明,体量单一,在城市天际线中增添了一个神秘的形状。该项目并不会打破马戏团原有的文化特征和集体形象,而是在某些技术和建筑方面进行了创新:帐篷外围的帆布、褶皱和颜色根据表演来进行变化,一个圆形走廊内分布着入口和出口。马戏团帐篷的奇妙之处在于其神秘感,以及其内的元素是如何轻易地转化为壮观的场景的。

运送设备的公共通道在演出当天布置好,形成一出"马戏团街"的剧目:一条充满欢乐氛围、笼罩在帆布之下的街道,观众和表演者都能融入到这种节日氛围中。

Circus Arts Conservatory

Located on the site of the Espagne barrack, the CCIR (Circus Center for Innovation and Research) is the first step in Auch's urban replanning. This program includes housing for embers of resident troupes, studios, a large rehearsal room, and a permanent circus tent, able to seat 700 people. This project introduces three different worlds, which, when put together, form a cohesive and readable space.

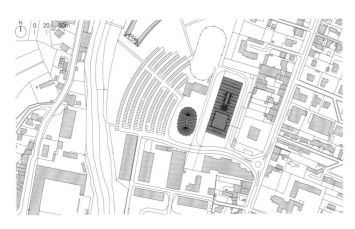

项目名称：Circus Arts Conservatory / 地点：Auch, France
建筑师：Doazan + Hirschberger & Associés / 主要建筑师：Benoîte Doazan, Nicolas Novello, Stéphane Hirschberger
织物装箱设计：Abaca / 音效设计：Altia / 木工、金属设计 & 木构建筑设计：Anglade / 经济学家：Atce
舞台装饰：Changement A Vue / 木材施工：Pyrénées Charpentes / 用地面积：4,100m² / 竣工时间：2012
摄影师：©Hervé Abbadie (courtesy of the architect)

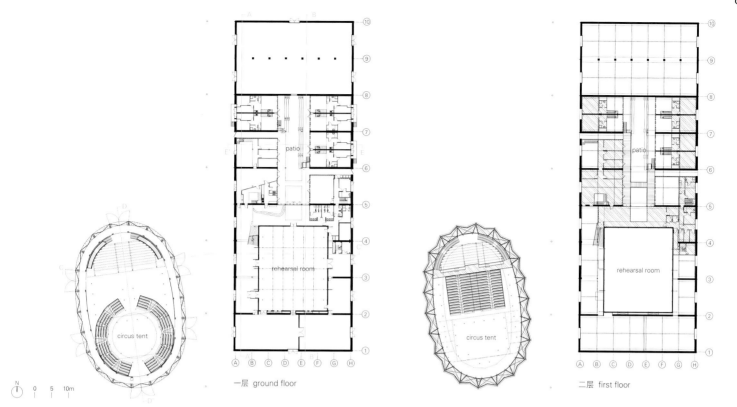

一层 ground floor 　　二层 first floor

First, the Patio, around which the different CCIR living spaces are organized. It is a half-courtyard, a space at the heart of the residents's life, in which they will be free to work, talk and dream.

The Atelier is then set all around the rehearsal room, and is the central workspace for circus artists: taking into account the high levels of intensity and concentration, this scenographic tower where performances occur is conceived as a convenient tool, the most effective possible. Its facades will be covered in larch, from which will appear converted stables.

Stationary, the circus tent, becomes a strong and recognizable urban beacon: opaque, and with a singular volume, it installs its enigmatic form in the city skyline. This project will not break away from the cultural and collective image of the circus, but rather innovate on certain technical and architectural aspects: a canvas shell, folds, colors that change according to the show, a circular gallery that even distributes entrances and exits. The magic of the circus tent is in its mystery, in how its work element easily transits into the spectacular.

Public access to the equipment is organized on show days, to form a sort of "circus street": a joyful and covered strada, where crowds and performers will be able to mingle in a festival spirit.

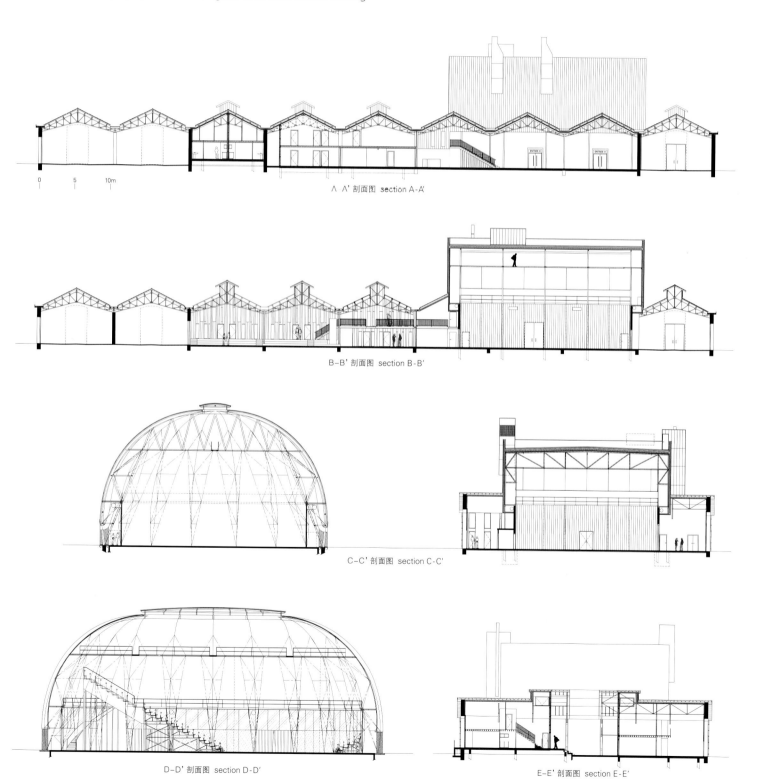

A-A' 剖面图 section A-A'

B-B' 剖面图 section B-B'

C-C' 剖面图 section C-C'

D-D' 剖面图 section D-D'

E-E' 剖面图 section E-E'

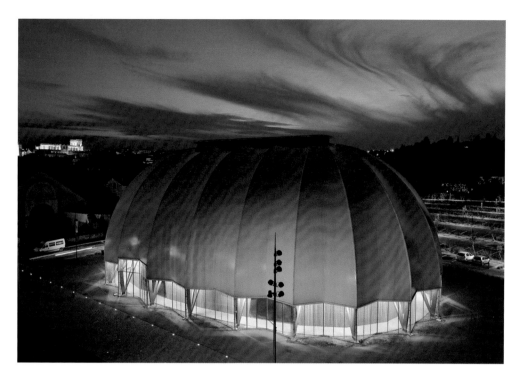

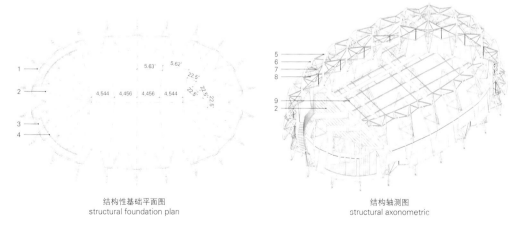

结构性基础平面图
structural foundation plan

结构轴测图
structural axonometric

1. triangulated posts BLC 180/180 2. BLC poles bleachers 3. stairs 60mm 4. holdings 100/100 + railing 100/260 5. upper crown shape IPE160
6. arc BLC 200/400 7. triangulations horiz. BM 120/120 8. lower crown shape IPE160 9. primary steel bridge IPE160 gateways

Allez-Up攀岩中心坐落于蒙特利尔西南区的拉欣运河旁，是一家攀岩健身馆。馆中新增的崭新的设施得益于一项重大的都市与社会改造工程。

1825年，拉欣运河开通，解决了船只在附近上游急流水域的通行问题。这条新兴的贸易之路开启了运河周边地区以及现如今的西南区的工业发展。19世纪60年代，该地区已成为一个繁忙的工业区，事实上，它是国内最为多元化的工业企业聚集区。在此期间，雷德帕斯糖果厂于1952年建造了4个筒仓作为仓库。然而在1970年，运河正式关闭，导致多家公司迁址，最终致使该地区被遗弃荒置。尽管这片区域的其他工业建筑逐渐都被重新利用或改造使用，但近40年来，这几个筒仓却被人们完全遗弃了。

Allez-Up攀岩馆的旧址位于附近的一处经过改造的旧建筑中。客户希望将健身馆扩充到原来的至少3倍大小，并且认为这处场地以及场地上的筒仓对于攀岩馆而言，能够发挥惊人的潜力。重新改造和利用废弃的筒仓，是以一种独特的方式来发挥蒙特利尔这些废弃的工业历史遗迹的潜能。在加拿大，这种从工业到娱乐用途的转型尚属首例，因此显著提高了拉欣运河地区在休闲娱乐、观光旅游方面的吸引力。

各个圆柱形体块之间通过一个矩形建筑结构相互连接，这座矩形建筑占据了建筑内部的大部分空阔场地。这种组合形式既在建筑内部重新划分了新的界限，同时也使四个废弃的筒仓相互连接。

建筑采用落地窗设计，为室内提供全天候的自然采光。立面的棱角结构与筒仓垂直、坚硬的质感形成有趣的对话效果，金属质地的建筑外观设计巧妙地呼应了周边地区的工业特色。每当夜幕降临，透过巨大的玻璃墙面人们可以从室外看到建筑内部的景致以及室内正在进行的精彩刺激的活动。

室内的攀岩墙粉刷成白色，象征着结晶的糖，提醒游客建筑的前身曾是雷德帕斯糖果厂的筒仓。

本案中的四个筒仓经过精心设计相互连接在一起：其中一个作为建筑的中心枢纽，仓内设置综合设施的主入口和通往其他筒仓的路径，其他三个筒仓则用于设置攀爬路线。

筒仓内部的顶端原先覆盖着两层大幅的雪松木板，用来防止存储的糖受潮，但是为了满足筒仓的功能转变，其中的一部分木板只能拆除。

建筑设计遵循生物气候学原理，所有的采光窗都朝向东南方，确保在冬季能够最大限度地吸收日光照射，获得热能，同时保证为建筑内部

Allez-Up攀岩中心
Smith Vigeant Architects

提供全天候的自然采光。此外，一些机械装置系统可将吸收的效率最大化，同时确保节能环保。在这样空旷的空间环境中，为使室内人员感觉舒适，很有必要设置一个地面辐射采暖系统。筒仓原有的细长形状也被设计师们充分利用作为烟囱，为建筑打造自然通风的效果。

砂糖传送带系统以及原先的机械系统等留存下来的部分工厂设施需要拆解和移除。留存下来的混凝土结构则需要加固才能适应新的楼层、洞口和室内人员的负载。

Allez-Up Climbing Center

Situated on the Lachine Canal in Montreal's Southwest borough, rock climbing gym Allez-Up's newest facilities are part of an important urban and social renewal project.

In 1825, the Lachine Canal opened as a solution to allow boats to bypass the nearby rapid upstream. This new path of commerce kick started the industrial development around the canal and the area of what is now the Southwest borough. By the 1860s the area was a busy industrial neighbourhood; in fact, it was the most diversified concentration of industrial establishments in the country. During this time, in 1952, four silos were construction by Redpath Sugar Refinery as storage for the factory. However in 1970, the canal was officially closed causing companies to relocate and eventually led to the desolation of the neighbourhood. While other industrial buildings in the area became reused and repurposed, for nearly 40 years, the silos were left completely abandoned.

Allez-Up's previous gym was located nearby in one of the repurposed buildings. The client wanted to increase his capacity by at least 3 times and believed that the site and the silos offered an amazing potential for a rock climbing gym. The repurposing of the abandoned silos was a unique way to exploit the potential of Montreal's abandoned industrial past. Their transformation into recreational use was the first intervention of its kind in Canada and has significantly added to the recreational and touristic attractions on the Lachine Canal.

The cylindrical volumes were thus connected by a rectangular form which infilled most of the vacant site. This reconfiguration created new limits within the site and simultaneously connected the abandoned sets of silos.

典型的建筑布局
typical building organization

垂直平面重新定义的空间
rethinking the space by vertical planes

折叠的平面作为攀爬表面
folding the planes for climbing surfaces

定义立面上复制的垂直平面
defining the vertical planes replicated on the facade

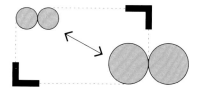

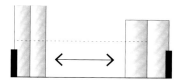

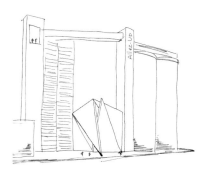

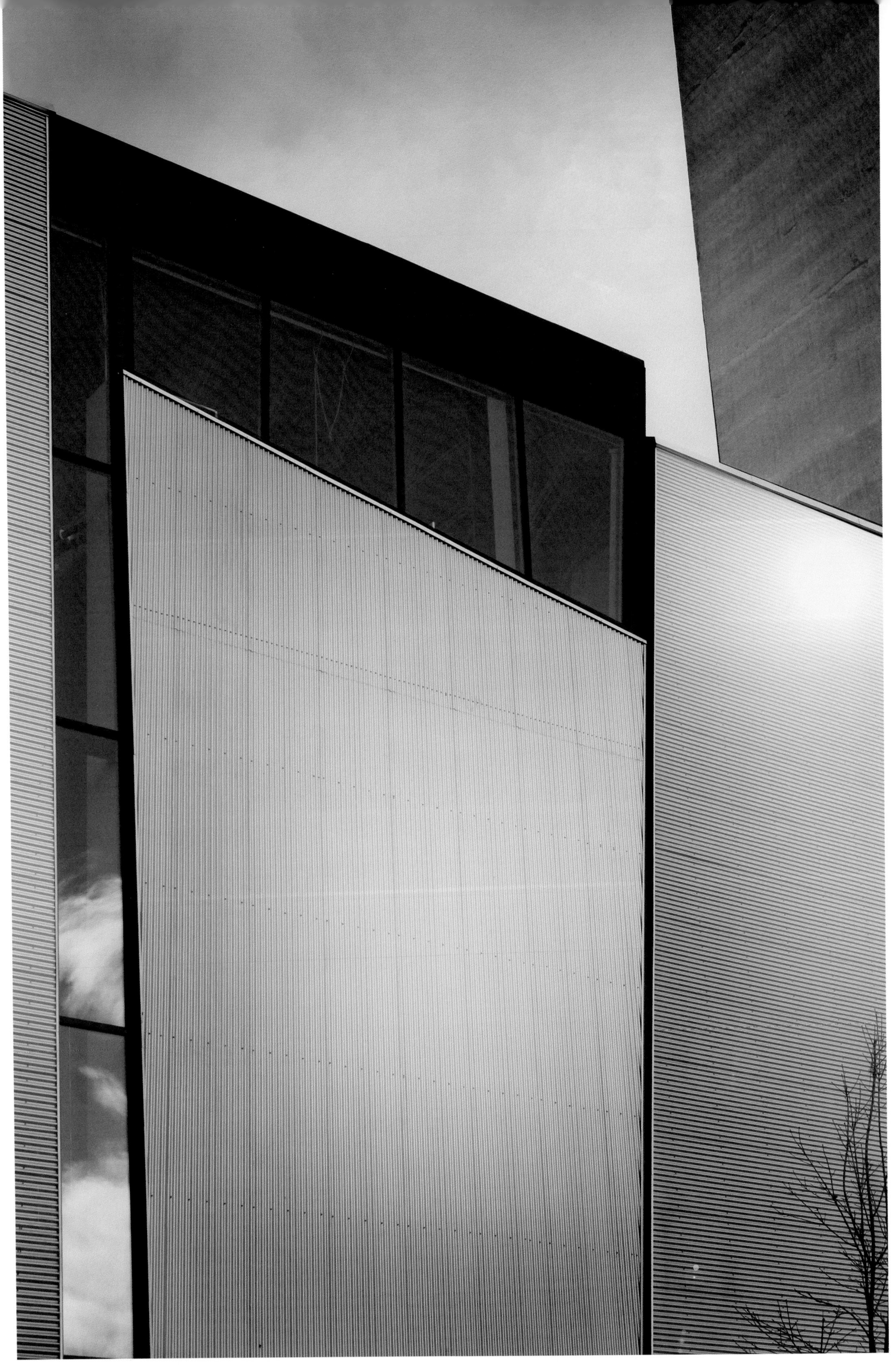

1. parapet(curtain wall)
   - glass spandrel panel
   - airspace 25mm, infilled with insulation
   - plywood 16mm
   - structure
   - plywood 16mm
   - extruded polystyrene insulation 38mm
   - roof membrane
2. parapet
3. galvanized steel flashing color: dark brown
4. underlayment membrane
5. wood blocking
6. insulation
7. wood blocking 2x3
8. steel shelf angle
9. glass spandrel panel
10. galvanized steel panel
11. corrugated steel panel
12. glass curtain wall
13. continuous sealant
14. galvanized steel plate attached to the purlin and to the beam, insure watertightness; continuous sealant at perimeter
15. steel beam
16. infill cavity with spray in place polyurethane insulation
17. continuous mastic sealing compound at perimeter
18. parapet(metal siding wall)
    - corrugated steel panel 22mm
    - galvanized finish
    - air barrier
    - galvanized z-bar 150mm
    - semi-rigid rock wool insulation 150mm
    - plywood 16mm
    - metal studs @405mm c/c
    - semi-rigid rock wool insulation 100mm
    - plywood 16mm
    - HSS
    - extruded polystyrene 90mm
    - roof membrane
19. "c" channel
20. galvanized u-bar
21. sealant
22. steel shelf angle
23. vapour barrier
24. acoustical sealant
25. thermo tape

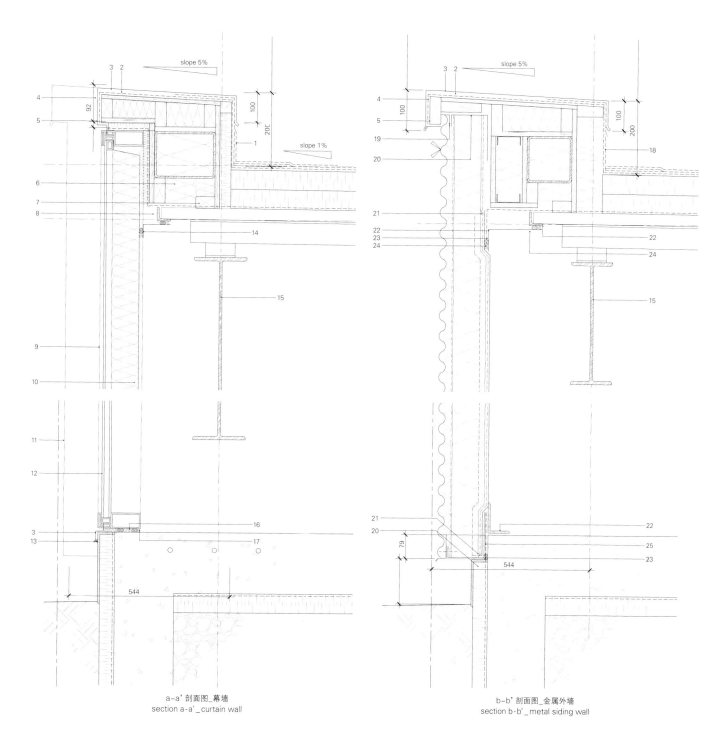

a-a' 剖面图_幕墙
section a-a'_curtain wall

b-b' 剖面图_金属外墙
section b-b'_metal siding wall

生物气候学设计原理	Bioclimatic design principles	可持续发展
自然通风/烟囱效应	natural ventilation / chimney effect	– 污染场地的修复
自然采光	natural light	– 减少热岛效应(100%白色屋顶膜)
日照得热量	solar gains	– 既有建筑的重新利用(95%)
		– 施工垃圾管理(90%转化率)

生态材料 Ecological materials
低有机挥发物饰面和材料 low voc finishes and materials
来自拆除建筑的回收和再生材料 reclaimed and recycled materials from demolition

Sustainable development
- rehabilitation of a contaminated site
- reduced heat island effect (100% white roof membrane)
- adapted reuse of existing structures (95%)
- construction waste management (90% diverted)

热质量 Thermal qualities
地面辐射采暖 radiant floor heating
温暖舒适区 thermal comfort zone

暴雨管理 Storm water management
用水管理及 water management & control
30%节约用水控制 30% water use reduction

能效表现 Energy performance
能量库 energy bank
优化能源性能 optimized energy performance
(持续节能40%的、符合美国采暖、制冷与空调工程师学会标准的建筑) (40% cont. reduction ASHRAE reference bldg)
地热连接 geothermal energy hook up

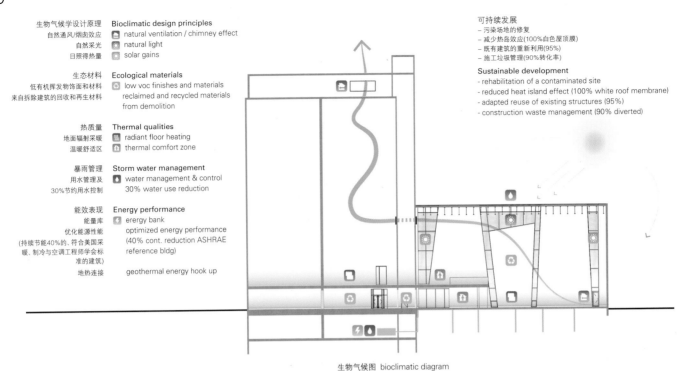

生物气候图 bioclimatic diagram

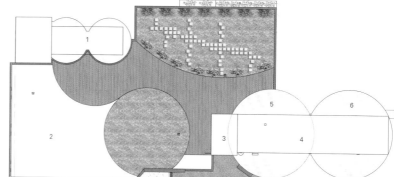

1 西侧筒仓	1. west silos
2 白色屋顶	2. white roofing
3 楼梯&斜屋顶	3. stairs & pitched roof
4 东侧筒仓的斜屋顶	4. pitched roof of east silo
5 筒仓屋顶1	5. silo roof 1
6 筒仓屋顶2	6. silo roof 2

屋顶 roof

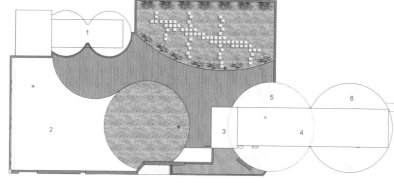

1 西塔楼	1. west tower
2 西侧筒仓	2. west silos
3 抱石运动区	3. bouldering
4 儿童及初学者攀岩壁	4. kids & beginners climbing
5 东塔楼	5. east tower
6 休息室和筒仓攀岩壁	6. lounge and silo climbing
7 东侧筒仓2期	7. east silo phase 2

二层 first floor

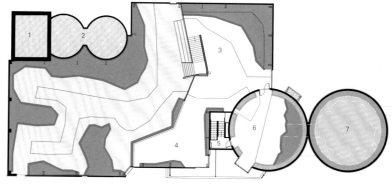

1 主入口	1. main entrance
2 筒仓外侧攀岩壁	2. exterior climbing on silo
3 西塔楼	3. west tower
4 西侧筒仓	4. west silos
5 铅质攀岩壁	5. lead climbing walls
6 攀岩大厅	6. rock climbing hall
7 抱石运动区	7. bouldering
8 瑜伽室	8. yoga room
9 东塔楼	9. east tower
10 接待处	10. reception
11 东侧筒仓2期	11. east silo phase 2
12 行政办公区	12. administration

一层 ground floor

The windows run full height and saturate the interior space with natural light throughout the day. The angularity of the facade provides an interesting dialogue with the verticality and solidity of the silos, while the metallic exterior finish nicely compliments the industrial character of the neighbourhood. At night, the large openings reveal the interiors and all its exciting activities taking place.

The climbing walls were painted white to symbolize crystallized sugar and to remind visitors of its previous function as the Redpath Sugar Silos.

The silos were carefully integrated in the program: one of them is acting as a hub with the main access and distribution of the complex, while the others are used for setting climbing routes.

The inside of the silos was originally covered with two layers of cedar wood planks running full height, originally used to protect the stored sugar from humidity. However in order to convert the silos some of them had to be removed.

Respecting bioclimatic principles, all the openings were oriented south-east to optimize solar heat gain during the winter and also ensure natural light throughout the day. The mechanical systems were developed to maximize their efficiency and ensure energy savings. A radiant floor heating system was necessary in ensuring the comfort of occupants within such a vast space. The inherent elongated shapes of the silos were taken advantage of and act as a chimney to create natural ventilation for the building.

The existing factory components had to be disassembled and removed, such as the sugar conveyor belt system and the original mechanical systems. The existing concrete structure had to be strengthened to accommodate new floors, openings, and occupant loads.

项目名称：Allez-Up Rock Climbing Gym
地点：Rue Saint-Patrick, Montreal, QC, Canada
建筑师：Smith Vigeant Architects / 总建筑师：Daniel Smith
设计团队：Daniel Smith, Karine Renaud, Anik Malderis, Étienne Penault, Cindy Neveu, Mélanie Quesnel, Stéphan Vigeant
工程师：NCK Inc & Martin Roy and associates
景观建筑师：Groupe Rousseau Lefebvre
总承包商：eSpace Construction Inc / 甲方：Richer – de la Plante Family
照明设计：Smith Vigeant architects, Martin Roy et associates
用地面积：1,990m² / 建筑面积：1,220m² / 有效楼层面积：1,525m² / 竣工时间：2013
摄影师：
Courtesy of the architect-p.76, p.83
©Stephane Brugger (courtesy of the architect)-p.74, p.77, p.78-79, p.80, p.81, p.82

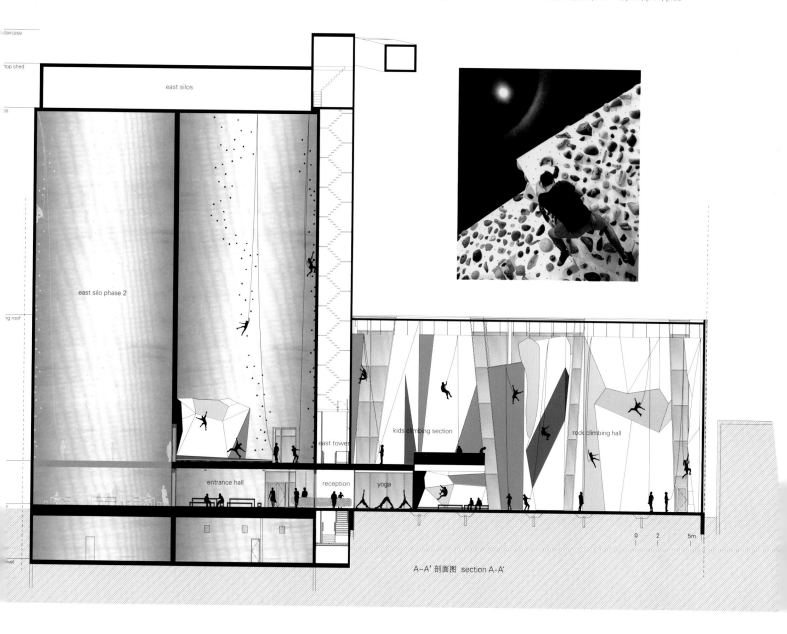

A~A' 剖面图 section A-A'

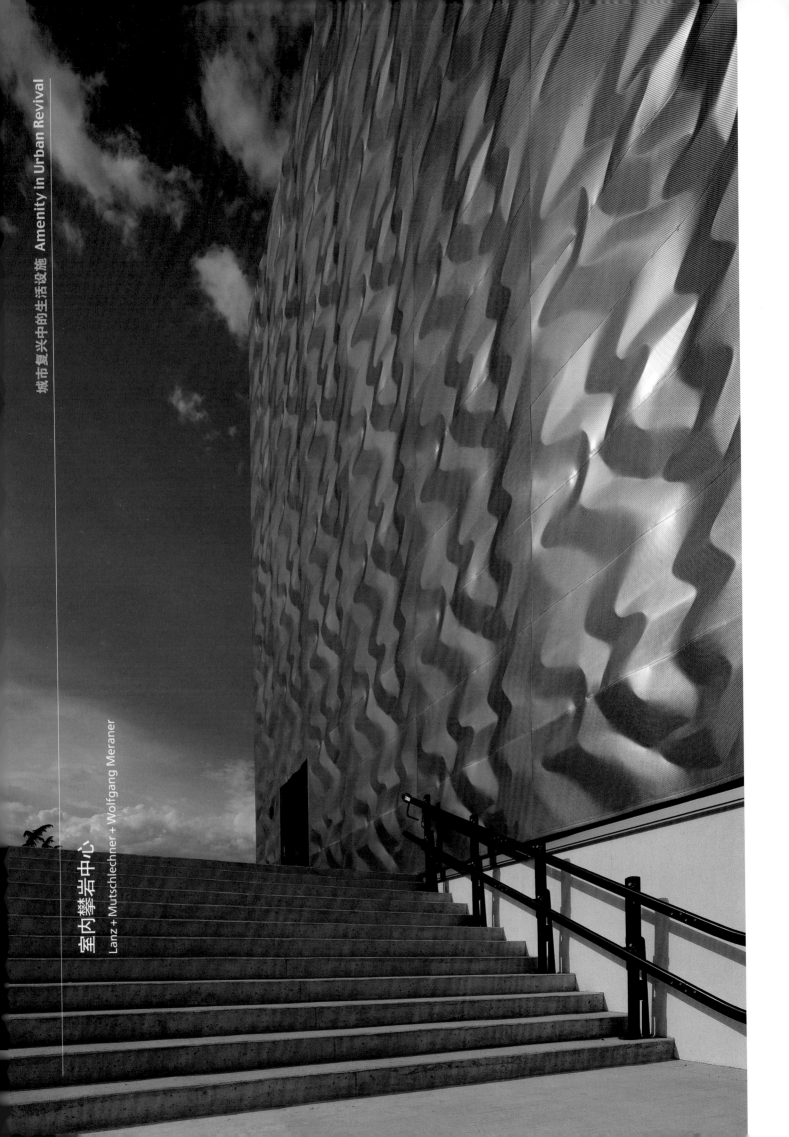

城市复兴中的生活设施 Amenity in Urban Revival

室内攀岩中心
Lanz + Mutschlechner + Wolfgang Meraner

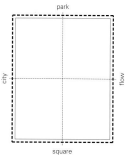

与周围环境的关系
relation with the surrounding

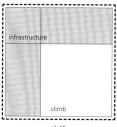

功能
program

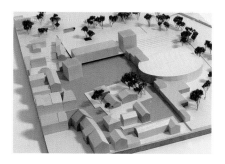

在考虑这座城市的城市规划的基础上，Bressanone攀岩健身房建在了一个显眼的地方。通过这种方式，该建筑博得一个高度的审美和设计标准。自然风景通过透明的立面进入室内，同时，人们从室外也可以看到室内的场景。多层的立面结构构成了一种波纹状的纹饰，从不同的角度看去具有不同的视觉效果；对室内攀岩的使用者来说，与外部的参观者一样，都与攀岩健身房形成了动态的关系。在建筑的规划中，生态和节能是基本的理念。这些都涉及了精细的规划和独特的多层立面气候理念以及空调和蓄热系统，建筑立面结构的中间层和蓄热体可以进行通风和蓄热。由于建筑中应用了太阳能、随季节变化的表面和自然通风系统，所以机械冷却是完全没有必要的，因此节省了建筑的建造和维护费用。人们从南侧的新城市广场和北侧位置较高的城市公园均可到达攀岩健身房的入口。新建的地下停车场将与健身房相连。健身房对伤残人士设有无障碍通道，还可用于治疗目的。开放的健身房结构设有分开的攀岩模块，人们在此可以欣赏到新公园和室外广场的风景，同时还能看到整个健身房的概观。这座建筑可满足专业和业余水平攀登者相应的攀岩需求，同时还可用于教学和比赛时使用。

Indoor Rock Climbing Center

The climbing gym of Bressanone was built on a prominent spot regarding the urbanism of the city and in this way, the building got ambitious aesthetics and design. Through a transparent facade, nature and landscape can be transported inside the gym and at the same time the inner happening of the building is visible from the outside. The multi-layered facade structure produces a moire effect, which allows new impressions for the observer all the time and relates the user inside and the observer outside in a dynamic relationship to the climbing gym. During the planning of the building, ecology and sustainability were basic ideas. They are involved through a precise planning and a specific climate concept with a multi-layered facade, air conditioning and accumulation of heat,

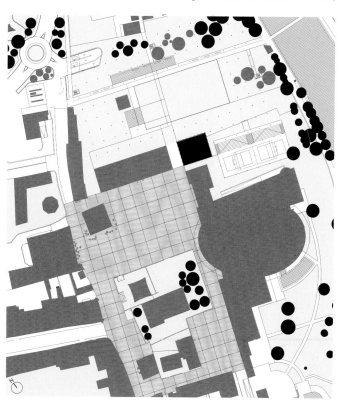

室内攀岩墙体的设计理念
indoor climbing walls concept

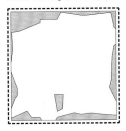  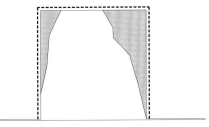

自然照明理念
natural lighting concept

 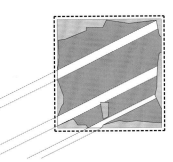 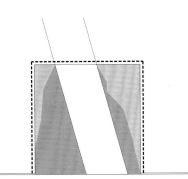

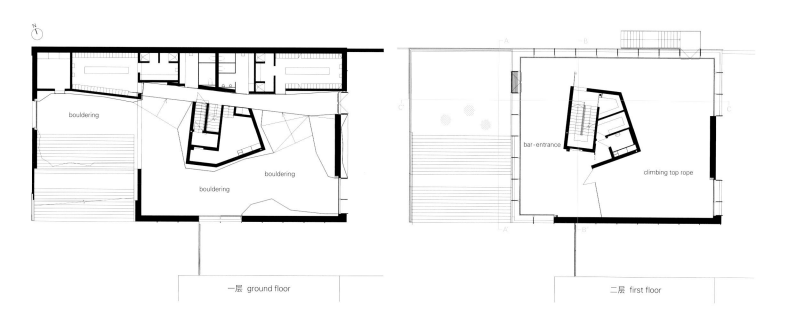

| 一层 ground floor | 二层 first floor |

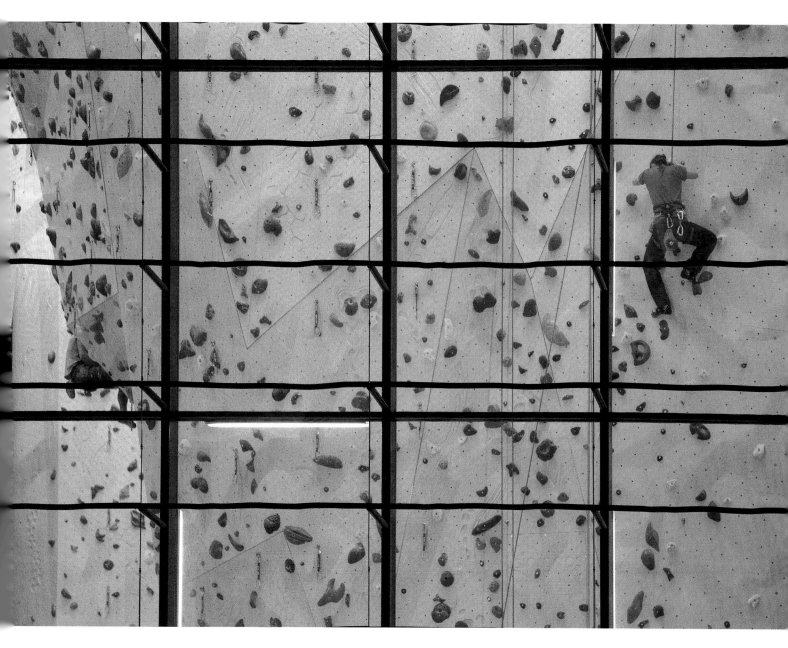

provided by interfaces in the structure of the facade and thermal mass in the building. There is no need for mechanical cooling by using the energy of the sun, the seasonal changing of the facade and the natural ventilation, which result in saving costs for the construction and the maintenance of the building. The entrance of the climbing gym is accessible from the new city square in the south and from the higher located city park in the north. The new underground parking will be connected to the gym and for the handicapped accessibility, the gym can also be used for therapeutic purpose. The open gym structure with the detached climbing blocks allows a view to the new park and the square outside and gives an overview of the whole gym at the same time. The climbing requirements correspond to the level of professional and amateur climbers, as well as for instructions and competitions.

项目名称：Vertikale Indoor Rock Climbing
地点：Bressanone, Italy
建筑师：Lanz + Mutschlechner, Wolfgang Meraner
施工单位：Domus
立面：Frener & Reifer
结构工程师：Tragwerksplanung Dr. Ing. Andreas Erlacher
室内攀岩墙体设计：Sintgrips
暖通空调系统设计：Energytech, Herman Heiztechnik
电气工程师：Euro Impianti Elettrici Bolzano
屋顶设计：Fischnaller J.
木匠：Castiglioni Gitzl
用地面积：8,000m²
总建筑面积：800m²
有效楼层面积：325m²
施工时间：2009—2012
摄影师：©Günter Richard Wett(courtesy of the architect)

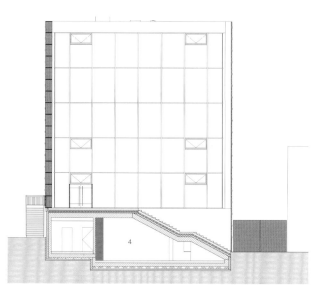

A-A' 剖面图 section A-A'

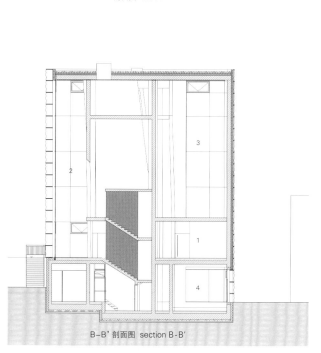

B-B' 剖面图 section B-B'

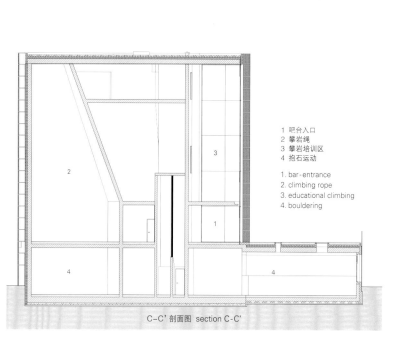

C-C' 剖面图 section C-C'

1 吧台入口
2 攀岩绳
3 攀岩培训区
4 抱石运动

1. bar-entrance
2. climbing rope
3. educational climbing
4. bouldering

城市住宅：
空间关系—地域性—隐私性
Urban Dwell:
Proxemics - Territo

全球化促进了全球用户文化的发展。在这样的文化中，无论是消极的还是积极的，各国的需求都有着相似的联系。尽管如此，文化代码还是深深地影响了全球用户的生活方式。地域性和个人隐私便是这种独特的人类研究带来的直接结果，这种研究被人们称作"人际距离学"或者"符号学"。符号的象征和象似性由文化习俗决定，并且互相联系。文化习俗对符号的影响越深，那么符号获得的象征价值也就越高。Claude Raffestin推测，地域性是由社会—空间—时间构成的三维背景下的人类和环境之间的抽象或具体的关系。这是一个开放的过程，它取决于人类及其与空间之间建立的关系种类，这个观点与Robert D. Sack教授提出的理论相反，他认为地域性与社会力量的控制及其主要表现相关。在建成环境中，占主导地位的室内和室外空间便是这一理论的直接结果，也就是人们熟知的"生活小区"。这些报告可以对室内产生不同的管理方式，而管理方式影响了建筑的外观及最终形式。内向型房屋和空间分层性影响了城市规划，产生了城市矩阵式的增长。城市绿化空间占据重要作用，它与个人隐私的概念紧密相连，一方面继续保持客观物体与背景之间的对比，同时又与内在空间合并。

Globalization has caused the development of a global user culture where needs have similar associations, whether positive or negative, across numerous countries. Despite of that, culture codes still strongly influence the lifestyle of the people around the world. Territoriality and privacy are two direct results of particular human studies, better known as "proxemics" and "semiotics". The iconicity and symbolism of a sign depend on the cultural convention and, are on that ground in relation with each other. If the cultural convention has greater influence on the sign, the sign gets more symbolic value. Claude Raffestin has postulated that territoriality is a result of the relationships(real or abstract) between man and environment in a tridimensional context formed by society-space-time. It is an open process that depends on the people and the type of relations established with the space. This vision is opposed to the theory affirmed by prof. Robert D. Sack, who instead said that territoriality is associated with the control and primary expression of social power. The dominance of the indoor and outdoor spaces is a direct result of this theory in the built environment, known as the human "biotope". These reports can generate a different management of the views inside the home, that influences the appearance of the building and its final form. Introverted house and spatial hierarchy affect the town plan, sometimes creating an urban matrix of growing. Emphasis is given to the role of urban green spaces, linked to the concept of privacy: on the one hand, it continues the object-background contrast, while, on the other hand, it merges with the interior space inhabited.

红房子_Red House/ISON Architects
SG住宅_SG House/Tuttiarchitetti
沙里夫哈住宅_Sharifi-ha House/Nextoffice
CM住宅_House CM/Bruno Vanbesien + Christophe Meersman
Namly住宅_Namly House/Chang Architects
西原墙_The Wall of Nishihara/Sabaoarch
津田沼住宅_House in Tsudanuma/Fuse-Atelier
1014住宅_House 1014/Harquitectes

城市住宅：空间关系—地域性—隐私性_Urban Dwell: Proxemics-Territoriality-Privacy/Fabrizio Aimar

riality - Privacy

如今，人类处于创造整个世界的位置，生态学家将其称为"生活小区"。[1]人类对规划空间形成了互相联系的结论和理论，它们被称作"人类距离学"。这种分类系统背后的假设与人类的天性有关，它展示了这样一种我们常常把它称作"地域性"的行为，人们据此来区分空间与空间之间的距离，做的每一件事都与空间体验有关。针对于人类距离学，在这个意义上最有趣的方法由微型文化区开展，它包括三个方面，包括固定特征、半固定特征和非正式特征。尤其，固定特征空间是组织个体和群体活动的最基本的方法之一，在下面的分析和个例研究中会提到。说到Marc Fried的推断，"……家是这样的一个局部地区，在这里你可以体验到生命中最有意义的一些方面……"[2]，它与固定特征空间之间的关联很清晰。实际上，目前那些欧美人认为理所当然的房屋内部布局，是最近才发生改变的，也就是说很近代。直到18世纪，欧洲家庭的房屋才有固定的功能[3]。如今，三口之家是没有私人空间的，因为没有设计专门的空间用作某一个特殊功能。从那以后，这种家庭模式便固定下来了，这种改变进一步体现在家庭式的房屋里。

下面的个例研究清晰地表明了个人隐私这个概念的不同方面在房屋本身以及外围的构造中的体现。第一种情况对空间进行了严格的划分层级，产生了一种对人类室内活动的控制，如果有人从第二点，也就是私密的外向型来分析的话，那么论述就会变得更有意思了。它通常导致一种形式的内向性，与建成的框架相比，它与主导型关系紧密相连。这种选择可以作为感应型下层住宅高档化过程的设计工具，比如，通过使用小型窗户，可以限制与城市环境比较下的视觉关联。这种亲密概念的变化形式在某些存在个人项目的地区中表现得也很明显。欧洲、亚洲和中东之间都有很大的区别。在美国，第一部法律评论性文章《隐私权》[4] 于1890年发表，里面用了三段来说明生活和财产的独立权。相反，日本没有类似隐私这样的相关词汇[5]，但是这并不能说明日本人之间就不存在个人隐私的概

Today, mankind is in the position of creating almost the entire world in which it will live, what the ethologists refer to as "biotope".[1] The interrelated observations and theories of man's use of these planned spaces are defined by the term "proxemics". The hypothesis behind this classification system is related to the human nature, that exhibits a behavior which we commonly call territoriality. According to this, man uses the senses to distinguish between one space or distance and another. Everything that he does is associated with the spatial experience. Focusing on proxemics, the most interesting approach in this sense is carried on by microculture, that includes three aspects: fixed-feature, semi fixed-feature and informal. In particular, the fixed-feature space is one of the basic ways of organizing the activities of individuals and groups, mentioned in the following analysis and case-studies. Referring to the Marc Fried's postulate "[…] home is […] a local area in which some of the most meaningful aspects of life are experienced."[2], the relevancy of the connection to fixed-feature space is clear. Actually the present internal layout of the house, which Americans and Europeans take for granted, is quite recent. Rooms had no fixed functions in European houses until the eighteenth century.[3] Members of the family had no privacy as we know it today, because there were no spaces specialized for a particular type of use. From that period, the family pattern began to stabilize and this change expressed itself further in a domestic form of the house.

The following case-studies clearly show the different shades of the concept of privacy, understood both within the house itself and externally. The first condition generates a control of the activities that takes place within, setting a strict hierarchy of spaces. The discourse becomes more interesting if one analyzes, instead, the second point above: the privacy outwards. It leads to a formal introversion, in close connection to dominance/dominated relationships compared with the built context. This choice is often a design tool of an induced gentrification process, such as, for example, the limitation of visual relations compared with urban environment by the use of a few smaller windows. Variants to the concept of intimacy are also evident by region in which individual projects are located. There are many differences between Europe, Asia and the Middle East. In the U.S.A., "The Right to Privacy"[4], the first law review article published in 1890, contains three paragraphs to describe "the right to be let alone" with regard to life and property. Conversely, there is no Japanese word for privacy[5]. Yet one cannot say that the concept of privacy does not exist among the Japanese

SG住宅，卡塔尼亚，意大利
SG House in Catania, Italy

念，而是说日本人关于隐私的概念和西方不同而已。他们把他们的房子和房子周围的地带看做一个结构体。在内部空间的使用方面，我们发现日本人的激进行为发生了很大的变化。在最近过去的几年中，他们总是使他们房间的任何一边都很干净，因为他们都是在中间的空间活动。相对的，欧洲人喜欢在墙边放一些家具来填补周边的空间。所以，西方的房间看起来总是没有日本人的那么杂乱。相反，阿拉伯人希望自己的房间能有很大的空间。阿拉伯中产阶级的房子的内部空间就很宽阔，他们不喜欢设置很多隔间，因为他们不喜欢孤独的感觉，他们设计了充足的无障碍空间来自由活动（超过100m²），天花板也很高，这样不影响室内的视野范围，而望向室外的视野也没有障碍。

此外还有一个很重要的问题，即将绿化表面和斜面作为附加值。我们可以认为这在美学方面或者概念上具有隐含意义（前花园或庭院），更或者说是天气方面的需求（缓解"热岛效应"）。第一个因素形成了城市发展矩阵，它确定了一种建筑风格，该风格也影响着城市规划。第二个因素是气候方面的要求，能够隔离热量，聚集热量，收集雨水用于灌溉以及自然通风。尤其需要注意像庭院这种缓冲空间的出现，这些地方使太阳能在狭窄和细长的地方也能接收到顶层的光线，同时，对主要空间和次要空间都能在层次方面进行控制，这非常接近内向型设计方法。

东永熙是韩国的一个农村，那里居住着大多数20世纪60年代保守的中高层阶级人群。就像个体生态学所说的，相对于地位比较低下的社会阶层，占主导地位的社会阶层更倾向于建立较大的私人空间。这种态度扎根于认知心理学、进化生物学、行为生态学、人类学和社会生态学中。在韩国，人们希望房子的前面有一个大花园，这种典型的布局决定了农村统一的造型，像一个整齐的矩阵图。关于房屋，主人希望每个孩子都能有自己的安全的独立空间。这些关于控制性和安静氛围的要求鼓舞了主人的隐私空间不被打扰，在家工作。像上面所说的，在18世纪之前，关于西

but only that it is very different from the Western conception. They consider their own house and the zone immediately surrounding it as one structure. In this use of interior space, we can note a progressive change of the radical behaviors of the Japanese. In the recent past, they usually kept the edges of their rooms clear because everything takes place in the middle. Contrarily, Europeans tend to fill up the edges by placing furniture near or against walls. As a consequence, Western rooms often appear less cluttered to the Japanese than they do to us. The Arabs, instead, dream of large spaces for their own home. The spaces inside their upper middle-class homes are wide: they avoid partitions because they do not like the loneliness. There must be plenty of unobstructed space in which you are free to move (over 100 square meters), and high ceilings, so that they do not impinge on the visual field, and, also, an unobstructed view outwards.

Another significant issue is played by the green surface and its declensions, as a real added value. We may consider it as an aesthetic or perceptual connotation (front yard or courtyard) or as a climatic requirement (to mitigate the "heat island effect"). The first of these two considerations leads to an urban development matrix that identifies a building type that influences, in its turn, the town planning. The second one deals with the climatic requirements, as thermal insulation, thermal mass, stormwater collection for irrigation and natural ventilation. A particular focus should be given to the presence of mediated spaces such as courtyards. These ones allow the zenith light to ensure solar gains in narrow and elongated lots, and, at the same time, a hierarchical control of master and mastered spaces, which is really close to an introverted approach. YonHee-Dong, in Korea, is a village where a conservative upper-middle class from the late sixties lives. Just as in ethology, the human dominant class tends to establish larger personal distances towards those who occupy lower positions in the social hierarchy. This attitude has roots in cognitive psychology and evolutionary biology but also draws on behavioral ecology, anthropology, closely linked to sociobiology. In Korea, people dream about having a large garden at the front of the house, and this typical arrangement has determined the homogeneous shape of the village, as a matrix. About the house, the client wanted, at the same time, to give a room to each of his children and keep his own private room safe from them. These requirements of stillness and control were imposed to encourage the work at home. As reported above, before the XVIII century, the western concept of childhood and its

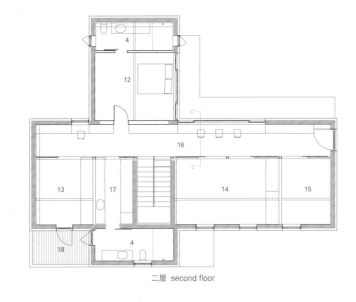
二层 second floor

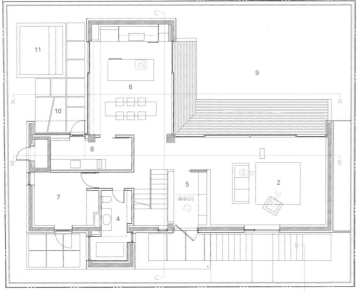
一层 first floor

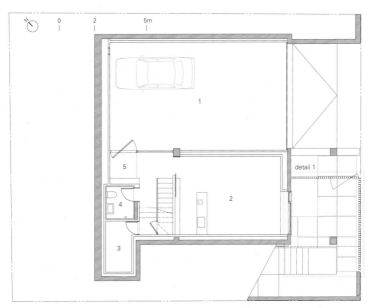
地下一层 first floor below ground

1 停车场	10 后院	1. parking lot	10. backyard
2 起居室	11 游泳池	2. living room	11. swimming pool
3 存储室	12 主卧室	3. storage	12. master bedroom
4 浴室	13 卧室1	4. bathroom	13. bedroom-1
5 入口	14 卧室2	5. entrance	14. bedroom-2
6 厨房/餐厅	15 卧室3	6. kitchen/dining	15. bedroom-3
7 客房	16 走廊	7. guest room	16. corridor
8 杂物房	17 前厅	8. utility room	17. front room
9 院子	18 露台	9. yard	18. terrace

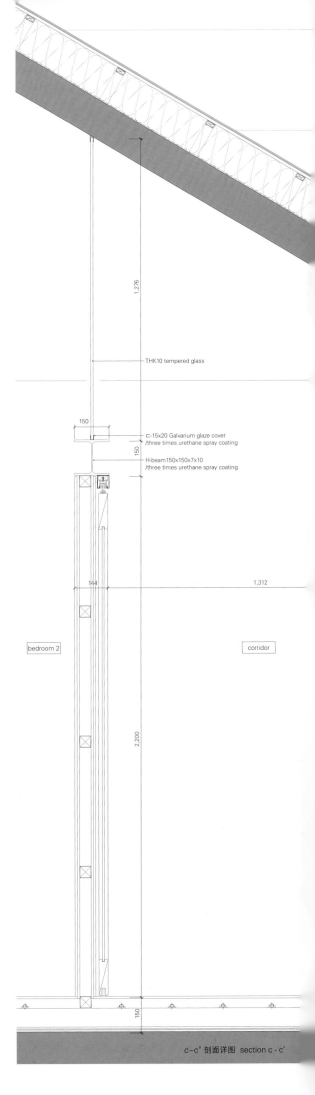

c-c' 剖面详图 section c - c'

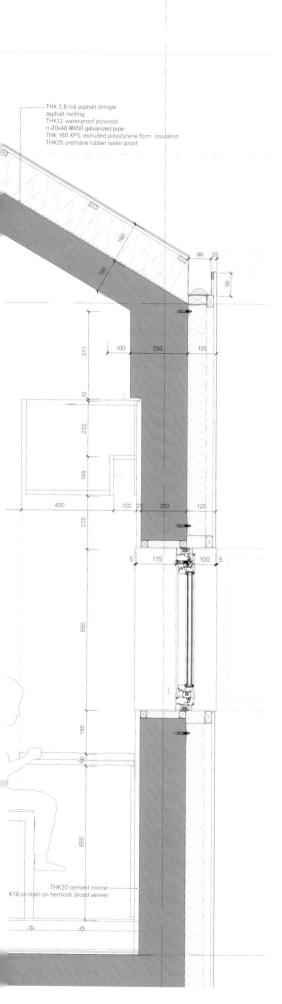

A-A' 剖面图 section A-A'

SG住宅
Tuttiarchitetti

建筑的一层宽5m，长20m，其大面积墙体采用熔岩砌成，位于成形的熔岩基础之上，这是20世纪70年代大规模建筑的遗物。该建筑只有成对角线分布的两个开口，东北开口朝向区域内的花园，西南开口可一瞥城市。一个多边形的木构小屋置于房屋的熔岩碎片之上，旋转6度并切除下部体量，对角线裂缝上的两个阳台使其成形。在周围耸立的高楼后方，它就像一颗精致的宝石一样存在。

该项目的改建工程包括扩建其高度，该建筑在20世纪初打算作为一个车间，然后再改建成住宅。现存的建筑是扩建的单层幸存建筑物，当时的扩建行动已经席卷了城市郊区的多家工厂。在特定情况下，用于构建周围高层建筑的土方工程，将熔岩露出地面部分（其上方矗立着现有的建筑）隔离开来，这些土方由熔岩砌砖组成，与地面融为一体，成为该项目的新基础平面。

新建筑位于旧建筑的顶部，木材被选定为建筑的材料，建筑师的这一决定不仅是因为其在节能及抗震方面实用性强等优点，而且还有凸显强制性嵌入结构本质的期望。

建筑师想要赋予木材的特性不仅由该结构体现出来，而且还从它由内到外的所有组件来展现；因此，建筑物整体都是由木材、木纤维保温材料和来自埃特纳火山的栗木表层构成的。栗木是当地一种古老而传统的建筑材料，展示出非凡的表现力，还有可观的技术品质。

建筑物的形状就是房子的原型，带有人字形的屋面；屋脊旋转了6度使屋面两边坡度不等，与其给人的熟悉感形成了潜在的抵触性。

SG House

A building at the ground floor of 5 meters wide and 20 meters long with generous walls of lava stone stands on a shaped core of lava, residue of massive 1970s' buildings, with only two openings diagonal, north-east towards the gardens of the district, south-west to a glimpse of the city. A prism wooden "hut" is placed on the lava fragment of the house, rotated 6 degrees and cut on the volume below, carved by two terraces on the diagonal gashes: a gem set between the heroic rears of tall buildings.

The project consists in expanding the elevation of a building of the early twentieth century already intended as a workshop and then adapted for habitation. The existing building, to a single elevation, is one of the surviving building expansion that has engulfed several factories on the outskirts of the city. In the specific case,

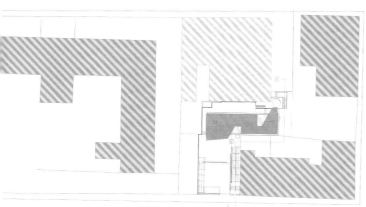

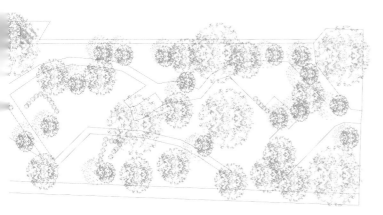

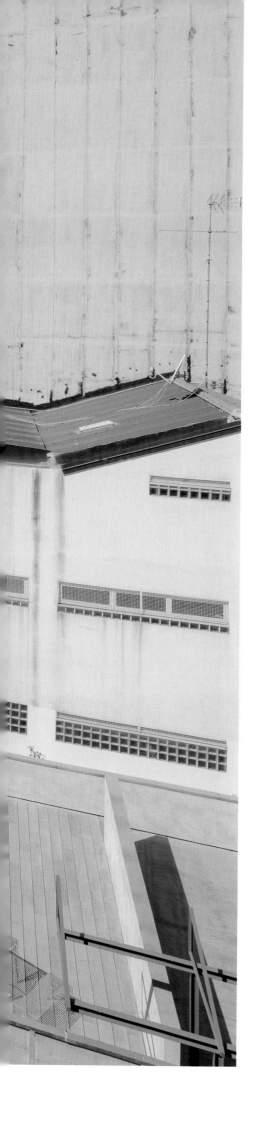

北立面 north elevation

南立面 south elevation

东立面 east elevation

0 1 2m

西立面 west elevation

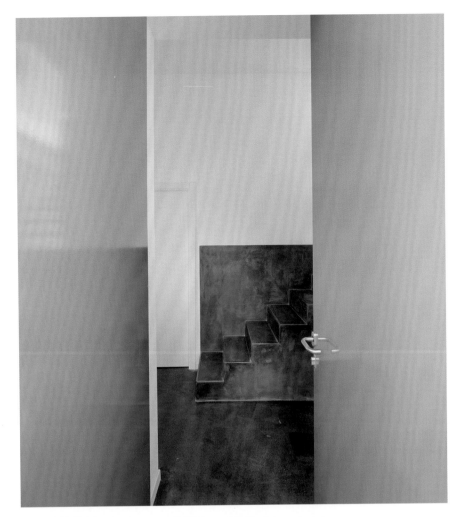

项目名称：Casa SG
地点：Catania, Italy
建筑师：Daniela Finocchiaro, Vincenzo Giusti, Luigi Pellegrino
总承包商：Salvatore La Spina & C.-Co.Ma.Ed. s.r.l.
结构工程师：Gabriele Correnti
用地面积：425.13m²
总建筑面积：133.15m²
有效楼层面积：259.25m²
结构：masonry lava stone_ground floor, wood(glulam cutted with CNC machine machine connected by stainless steel plates_upper floor
竣工时间：2014.9
摄影师：©Salvatore Gozzo (courtesy of the architect)

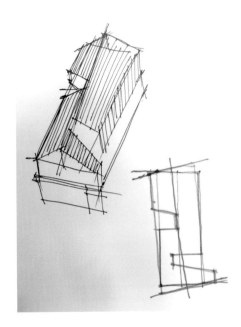

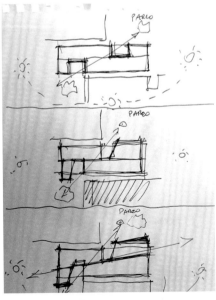

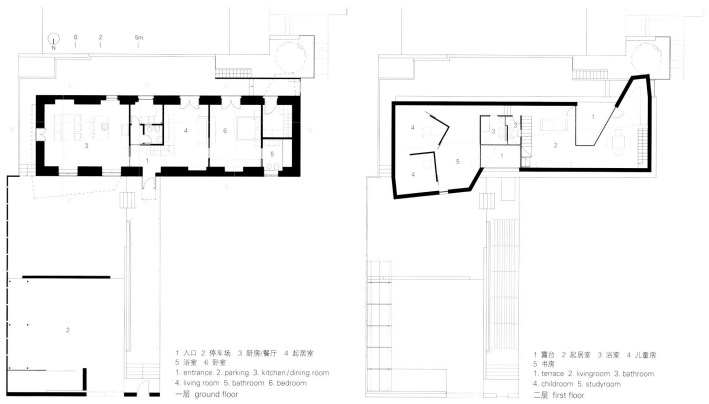

1 入口 2 停车场 3 厨房/餐厅 4 起居室
5 浴室 6 卧室
1. entrance 2. parking 3. kitchen/dining room
4. living room 5. bathroom 6. bedroom
一层 ground floor

1 露台 2 起居室 3 浴室 4 儿童房
5 书房
1. terrace 2. livingroom 3. bathroom
4. childroom 5. studyroom
二层 first floor

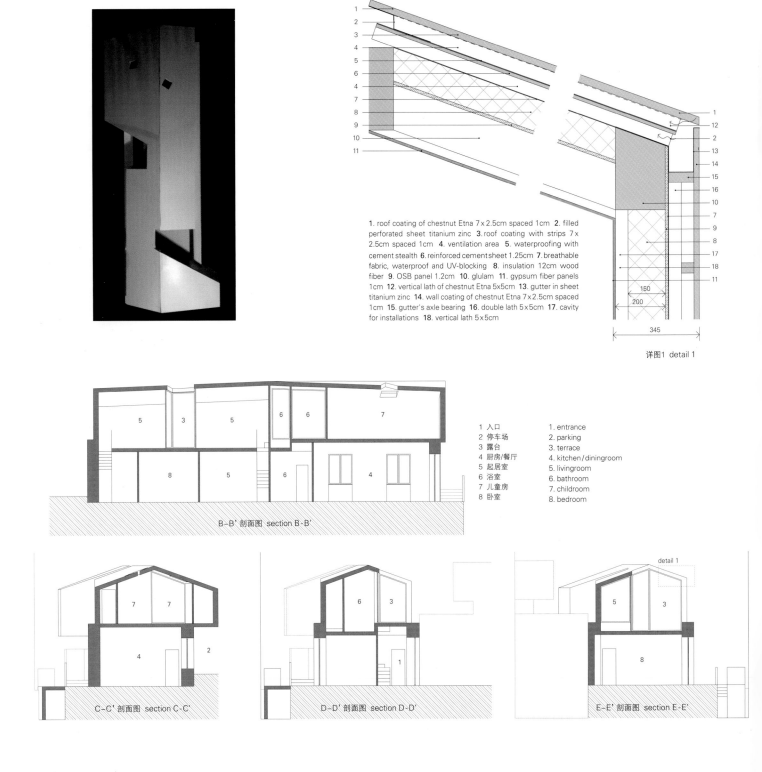

1. roof coating of chestnut Etna 7×2.5cm spaced 1cm 2. filled perforated sheet titanium zinc 3. roof coating with strips 7×2.5cm spaced 1cm 4. ventilation area 5. waterproofing with cement stealth 6. reinforced cement sheet 1.25cm 7. breathable fabric, waterproof and UV-blocking 8. insulation 12cm wood fiber 9. OSB panel 1.2cm 10. glulam 11. gypsum fiber panels 1cm 12. vertical lath of chestnut Etna 5×5cm 13. gutter in sheet titanium zinc 14. wall coating of chestnut Etna 7×2.5cm spaced 1cm 15. gutter's axle bearing 16. double lath 5×5cm 17. cavity for installations 18. vertical lath 5×5cm

详图1 detail 1

B-B' 剖面图 section B-B'

入口	1. entrance
停车场	2. parking
露台	3. terrace
厨房/餐厅	4. kitchen/diningroom
起居室	5. livingroom
浴室	6. bathroom
儿童房	7. childroom
卧室	8. bedroom

C-C' 剖面图 section C-C'

D-D' 剖面图 section D-D'

E-E' 剖面图 section E-E'

the earthworks, made to found the high buildings around, have isolated the lava outcrop on which stands the existing building; this, being composed of masonry lava stone, integrates with the ground and becomes the new foundation plan of the project.

For the structure of the new building, resting on top of the old building, was chosen the wood and that choice was dictated not only by practical advantages in terms of energy saving and resistance to earthquakes, but also from the desire to emphasize the nature of seditious intervention.

The character the architects wanted to give with the wood is not only manifested by the structure but from all of its components from the inside out, so the whole body of the building is made of wood, wood fiber insulation and coating chestnut from Etna, a material with an ancient tradition in local architecture, offering extraordinary expression as well as considerable technical qualities. The shape of the building is the archetype of the house, with the gable roof; but its familiarity is latently contradicted by the rotation of the ridge of 6 degrees.

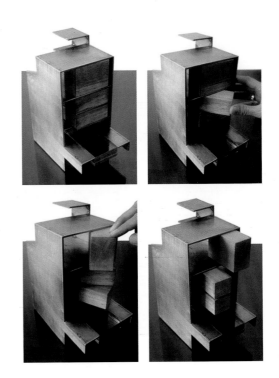

沙里夫哈住宅

Nextoffice

沙里夫哈住宅的核心设计理念是不确定性和灵活性。绝佳的室内空间品质和正规的外观构造，直接与建筑的旋转箱体结构的位置相呼应，将整座建筑打造成或开放，或封闭，集内向型与外向型特点于一身的建筑体量。而这些变化是根据季节的变化或楼层的功能场景而产生的。

和许多其他的城市地块一样，该项目场地内的立面宽度同其长度相比，显得非常狭窄。因此，建筑师必须将二维的立面转换为三维立面。在这里，建筑体量的开放型或封闭型特点参考了伊朗的传统住宅，根据季节的交替来改成动态的居住模式，为居民提供Zemestan-Neshin（冬季使用的客厅）和Taabestan-Neshin（夏季使用的客厅）。夏季，沙里夫哈住宅提供了一个开放且透明的穿孔体量，并设有宽大的阳台。相反，在德黑兰寒冷飘雪的冬季，建筑体块可以自行闭合，仅提供最少的洞口，也没有了夏季使用的宽敞阳台。在本项目中，对开放型和封闭型（内向型与外向型）概念的挑战成就了引人注目的空间变化，建造出一栋千变万化的住宅建筑。

建筑共分七层：下面两层是地下室，用作举办家庭宴会，放置健身器材和疗养的区域；一层用作车库和家政房。所有的公共活动都在二层和三层举行，家庭的私密生活区域位于四层和五层。

项目包括四个主体部分：分别是建筑结构的固定体量、上空空间、固定的体量以及可移动的体量。当旋转箱体结构处于闭合状态时，建筑就需要通过上空空间来获得采光，而这处上空空间同时也通过吊桥来连接两个固定的建筑体量。

该住宅能够满足居住者的功能需求。例如，业主可以根据是否有客人来访来改装客房（位于三层），以实现不同的使用目的。同样，也可以根据自身的喜好，来改变家庭办公室和餐厅（二层和四层的可旋转房间）的外观。换句话说，该建筑内部总是有可能出现不同的季节或照明的场景，其中的一些场景已经在项目的智能建筑管理系统程序中考虑到了。

在最初的设计步骤中，建筑师就已经注意到可以将室外的三棵松树融入室内空间。现在，当住宅处于开放的模式时，三棵松树与窗框恰好相映成趣。

在施工边界线后退3m左右的地方，建筑师建造了一个玻璃喷泉，为住宅的地下室提供了极好的日光照明。健身与疗养区设置在喷泉和地下水池之间，这里安装的反射装置可以将粼粼波光反射到整个空间。

Sharifi-ha House

Uncertainty and flexibility lie at the heart of the design concept in Sharifi-ha House. The sensational, spatial qualities of the interiors, as well as the formal configuration of its exterior, directly respond to the displacement of turning boxes that lead the building volume to become open or closed, obtaining introverted or extroverted character. These changes may occur according to changing seasons or functional scenarios of floor plans.

Like many other urban plots, the land for this project had a noticeably narrow facade-width compared with its length. Consequently, the expertise in transforming a two-dimensional facade to a three-dimensional one became indispensable. Here, the openness/closure of the building volume is a reference to traditional Iranian houses, which would dynamically serve as seasonal modes of habitation by offering both a Zemestan-Neshin (winter living room) and Taabestan-Neshin (summer living room) to their residents. In summertime, Sharifi-ha House offers an open/transparent/perforated volume with wide, large terraces. In contrast, during Tehran's cold, snowy winters the volume closes itself, offering minimal openings in total absence of those wide summer terraces. In this project, the challenges to the concepts of open/closed typology (introverted/extroverted character) led to an ex-

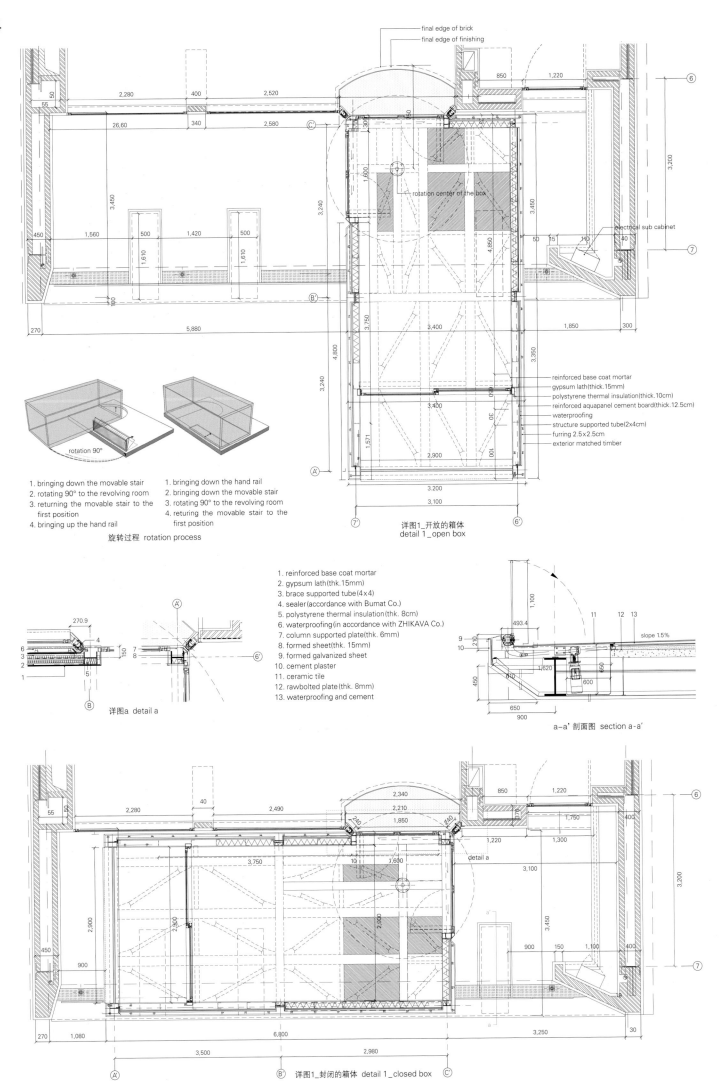

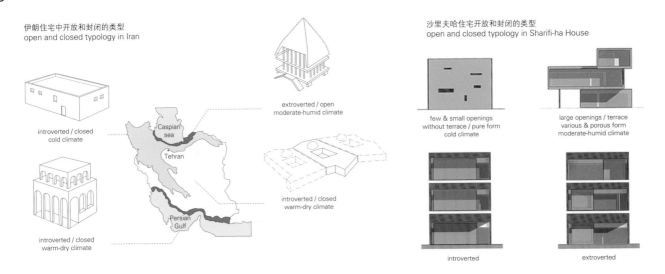

citing spatial transformation of an ever-changing residential building.

The house is distributed over seven floors: the two basement floors are allocated to family conviviality, fitness facilities, and wellness areas, while the ground floor hosts parking and housekeeping rooms. Public activities all happen on the first and second floors, and the family's private life takes place on the third and fourth floors.

The project consisted of four major parts; the fixed volume of the structure, the void, and the fixed volume and the mobile volume, respectively. When the turning boxes are closed, the building captures sunlight throughout the space of the central void, which also connects the two fixed volumes by suspended bridges.

The house adapts to the functional needs of its residents. For instance, depending on whether there is a guest or not, the guest room (located on the second floor) can be reconfigured for different purposes. Similarly, home office and breakfast room (turning rooms on the first and third floors) can change the formality of their appearance according to their residents' desire. In other words, there is always the possibility of having different seasonal or lighting scenarios, some of which have been already considered in the BMS program of the project.

From the initial design steps, the architects noticed that three pine trees outside could be incorporated into the spaces of the interior. Now, in the open mode of the house, the trees are pleasantly captured by the window frames.

Stepping back for about three meters from the construction boundary line allowed the architects to provide splendid daylight for the basement floors by inserting a glass fountain. The fitness and wellness areas are accommodated between this fountain and the basement pool where reflective installations reverberate the water's radiance all over the space.

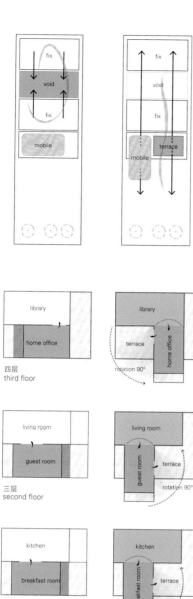

二层(开放)　　　　　　　五层
first floor(open)　　　　　fourth floor

一层　　　　　　　　　　四层(开放)
ground floor　　　　　　third floor(open)

1 桑拿房	1. sauna
2 淋浴室	2. shower room
3 冷水池	3. cold water pool
4 游泳池	4. swimming pool
5 按摩浴池	5. Jacuzzi
6 酒吧	6. bar
7 卫生间	7. w.c.
8 衣帽间	8. locker room
9 机械间	9. mechanical room
10 电梯	10. elevator
11 台球室	11. billiard hall
12 卧室	12. bed room
13 体育馆	13. gym
14 主要管道区	14. main duct
15 电气室	15. electrical room
16 家庭办公室	16. home office
17 入口	17. entrance
18 位于屋顶的水池	18. skylight pool
19 停车场	19. parking
20 门房	20. janitorial
21 存储室	21. storage
22 厨房	22. kitchen
23 餐厅	23. dinning room
24 主卧室	24. master bedroom
25 壁炉	25. fire place
26 起居室	26. living room
27 早餐室	27. breakfast room
28 露台	28. terrace
29 客房	29. guest room

地下二层　　　　　　地下一层　　　　　　三层(封闭)
second floor below ground　　first floor below ground　　second floor(closed)

128

项目名称：Sharifi-ha House
地点：No.3, Sharifi Dead-End, Saleh Hosseini St., Darrous, Tehran
建筑师：Alireza Taghaboni
设计事务所：Roohollah Rasouli, Farideh Aghamohammadi
细部设计事务所：Bahoor Office _ Hamid mohammadi, Amir Taleshi
细部高级顾问：Shahnaz Goharbakhsh
监工：Shahnaz Goharbakhsh, Alireza Taghaboni
项目合作：Mojtaba Moradi, Negar Rahnamazadeh, Asal Karami, Majid Jahangiri, Masoud Saghi, Hossein Naghavi, Fatemeh S.Tabatabaeian, Iman Jalilvand
施工单位：Imen Sazeh Fadak Consulting Eng
景观顾问：Babak Mostofi Sadri, Omid Abbas Fardi
结构设计：Sohrab Falahi / 机械顾问：Hoofar Esmaeili
电气顾问：Mohammad Torkamani / 旋转系统设计：Bumat Company
甲方：Mojgan Zare Nayeri, Farshad Sharifi Nikabadi
用地面积：407m² / 总建筑面积：238m² / 有效楼层面积：1,400m²
设计时间：2009 / 竣工时间：2013
摄影师：©Parham Taghioff (courtesy of the architect) (except as noted)

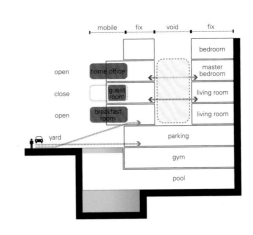

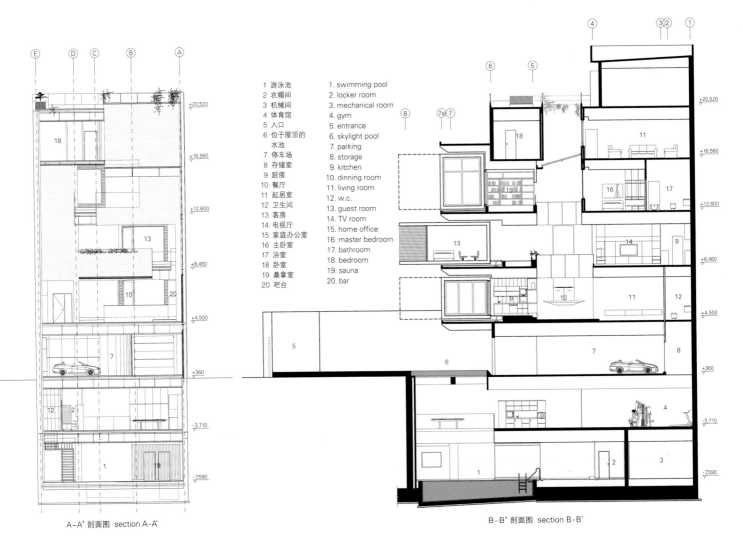

1. 游泳池 / 1. swimming pool
2. 衣帽间 / 2. locker room
3. 机械间 / 3. mechanical room
4. 体育馆 / 4. gym
5. 入口 / 5. entrance
6. 位于屋顶的水池 / 6. skylight pool
7. 停车场 / 7. parking
8. 存储室 / 8. storage
9. 厨房 / 9. kitchen
10. 餐厅 / 10. dinning room
11. 起居室 / 11. living room
12. 卫生间 / 12. w.c.
13. 客房 / 13. guest room
14. 电视厅 / 14. TV room
15. 家庭办公室 / 15. home office
16. 主卧室 / 16. master bedroom
17. 浴室 / 17. bathroom
18. 卧室 / 18. bedroom
19. 桑拿室 / 19. sauna
20. 吧台 / 20. bar

A-A' 剖面图 section A-A'   B-B' 剖面图 section B-B'

照片提供：©Majid Jahangiri(courtesy of the architect)

CM住宅

Bruno Vanbesien + Christophe Meersman

这栋新式住宅建在两处旧宅之间的空地上,两栋旧宅的历史可以追溯到20世纪的下半叶。别墅的主人是一名助理建筑师,这栋别墅就出自他和另一名建筑师之手,是他们成功合作的结果。

尽管从建筑的正面看上去,在其非洲红豆木质地的嵌板之后只有两层楼而已,但实际上这是一座要大得多的房子。建筑前部巨大的窗户,在一楼与二楼之间保持了稳定的平衡,也使本应清晰明了的建筑布局变得复杂化。在设计之初,设计团队就选择了一种分立式的建筑立面,使其能够与毗邻的住宅和谐相接。前门和车库的伪装设计进一步体现了这一点。前窗的设置不仅仅起到了美观作用。安置在基座上的大型窗户,既能保护一楼和二楼房间的隐私,这点是客户非常重视的,也使建筑本身成为闹市之中奢华的隐遁。

热带硬木制成的木质镶板被用来装饰建筑的正立面,也用来覆盖建筑的屋顶和背面,创造出了材料使用上和谐统一的效果。房屋的背面并没有使用同样谨慎、矜持的建造手法。在房屋的背面设计上,设计师选择追求最大程度的透明度与灵活性,自由表达房屋主人的个性特点。一楼和二楼设置巨幅的落地窗,窗户可以完全打开,这样屋内与屋外的界限就变得模糊,难以区分了。紧邻三楼卧室的露台淋漓尽致地体现了这种室内外界限模糊化的处理方式。那里的玻璃门可以滑动,使床可以滑到外面去,让人在星空之下安眠。玄武岩天然石材地板铺设在不同的房间以及外面的露台上,也有助于最大限度地实现这个效果。

置身室内,人们不会注意不到金属质地的楼梯。楼梯的独特设计将正上方采光井中的阳光最大程度地反射到了下方的楼层,保证了充足的采光。采光井的铝质框架隐藏在灰泥墙中,仅剩下玻璃在视野之内,使玻璃成为室内主要的装饰性元素。所有的内部元素(厨房、浴室、照明、家具)都是按照这一建筑风格来设计的。家具的严格布置、设计的创造性发挥、材料的精挑细选都很好地提升了建筑的风格,使其大放异彩。

House CM

This new house was built on an open lot in between two houses dating from the second half of the twentieth century. It is the result of the successful cooperation between the architect and the owner, an architect assistant himself.

Although the facade seems to suggest that behind its Afromosia paneling only two floors are hidden, in reality this is a far bigger house. The big window in front, evenly balanced between the two ground floor and the first floor, complicates an unambiguous reading of the layout even more. From the start, the design-team opted for a discrete facade, harmoniously linked with the neighboring houses. This can further be seen in the camouflaging of both the front door and the garage. The placing of the front window proves to be more than an aesthetically pleasing solution. The large window placed on a pedestal brings the rooms on both ground and

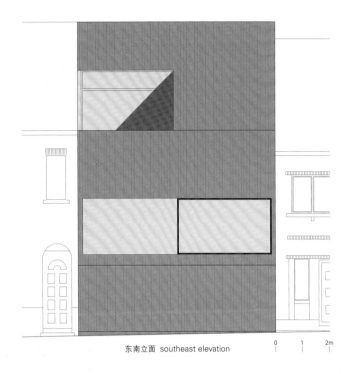

东南立面 southeast elevation

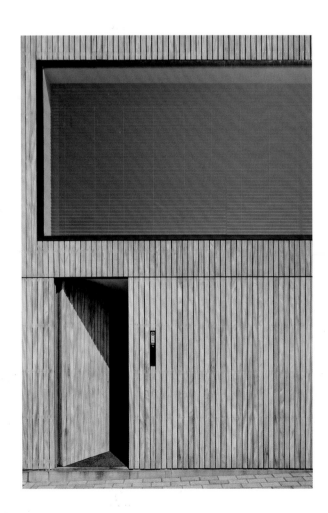

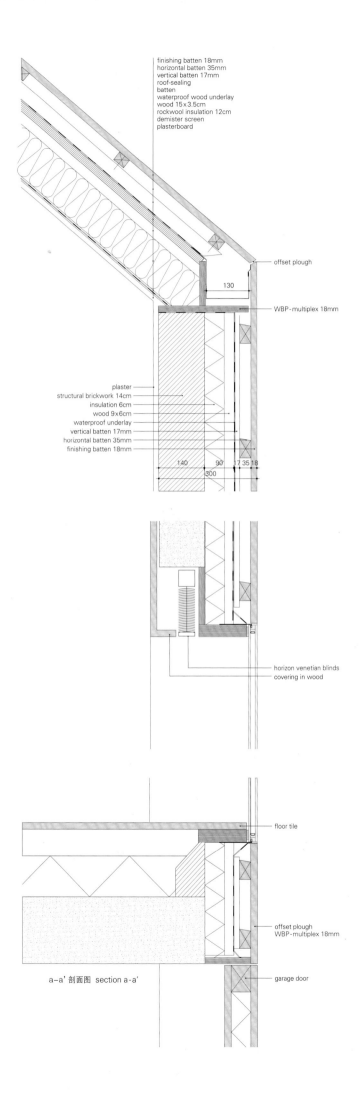

finishing batten 18mm
horizontal batten 35mm
vertical batten 17mm
roof-sealing
batten
waterproof wood underlay
wood 15 x 3.5cm
rockwool insulation 12cm
demister screen
plasterboard

offset plough

WBP-multiplex 18mm

plaster
structural brickwork 14cm
insulation 6cm
wood 9x6cm
waterproof underlay
vertical batten 17mm
horizontal batten 35mm
finishing batten 18mm

horizon venetian blinds
covering in wood

floor tile

offset plough
WBP-multiplex 18mm

garage door

a-a' 剖面图 section a-a'

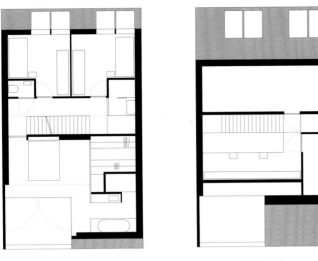

三层 second floor   四层 third floor

项目名称：House CM
地点：Zellik, Belgium
建筑师：Bruno Vanbesien, Christophe Meersman
项目建筑师：Christophe Meersman
用地面积：220m²
竣工时间：2011
摄影师：©Tim Van de Velde (courtesy of the architect)

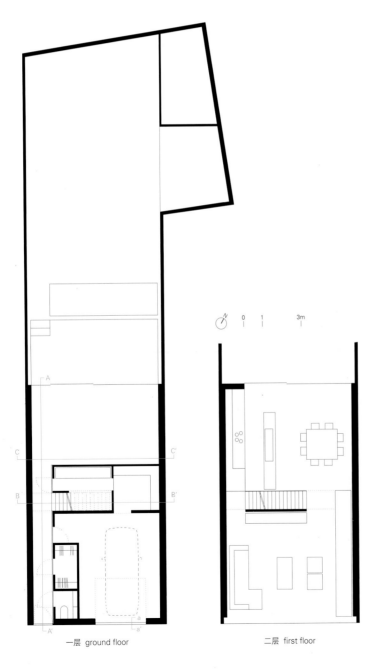

一层 ground floor   二层 first floor

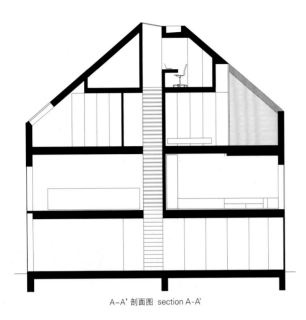

A-A' 剖面图 section A-A'

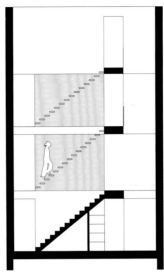

B-B' 剖面图 section B-B'

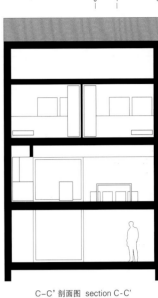

C-C' 剖面图 section C-C'

first floor a combination of much welcomed privacy and luxurious reclusion from the street.

The wooden paneling in tropical hardwood decorates the front facade but also covers the roof and the back of the house thus creating harmony and unity in the use of materials. The same discreetness and reservedness are not to be found in the back of the house. There the architect has chosen to pursue a maximum of transparency and flexibility and to give free expression to the personality of the owner. The large windows on the ground but also on the first floor can be opened up wide so that the transition between in/out becomes blurred. This playful treatment of in- and outside is best witnessed in the use of the terrace next to the bedroom on the second floor. There the glass door can slide away to make room for the bed which can be rolled outside to sleep under the stars. The natural stone floor composed of basalt tiles in the different rooms as well as the terraces outside maximizes this effect even more.

Inside one cannot but notice the metal staircase. Its construction is kept light to bring a maximum of light from the light shaft directly above the staircase to the stories below. The aluminum frame of the light shaft is hid between plaster so that only the glass remains visible making it a very decorative element in the house. All interior elements(kitchen, bathroom, lighting, furniture) were designed with the architecture in mind. The strict alignment of the furniture, the inventiveness of the design and the use of selected materials further accentuate the architecture and give it extra luster.

彼得和露西想要一个家,一个可以三代同堂的房子,这样他们既可以照顾孙子和孙女,又无需牺牲自由,在满足每代人不同的生活需求的同时,还能够保护个人隐私。彼得痴迷于混凝土建筑,概括来说他的设计要求就是"干净、简约、白色"。另外,他开列的设计愿望清单是"同一座房子但是能容纳两个家庭,简约但漂亮,有正面也要有背面,地处热带但是要凉爽通风,自然但不失品位,原始而又雅致,身居室内但是要与室外的大自然咫尺相连,小型但是宽敞,极简却绰绰有余,虽少犹多"。

祖父母的房间位于一层后方,既方便到达又非常隐秘,旁边紧挨着彼得的露天平台。这个露天平台是餐厅与客厅的空间延续,有利于促进三代人的交流。

父母的房间位于二层的最前端,这里可以隔过通风井瞭望双胞胎孩子的房间。这种布局实现了两代人之间隐私与和谐生活的平衡。

孩子们的房间在二层的中央,位于父母的房间和书房之间。这样对于孩子和父母来说,既能够相互看到也方便相互接触。这处空间在视野和通风上都有很好的优势,而且如果双胞胎长大后想要分开独立居住,也可以隔成两个房间。

将这些空间相连接起来的是一条贯穿前后的通道。这条通道是家庭聚会的地方,是孩子们最喜欢的游乐场,实际上也是东北和西南季风的风道。如果开车来访,经过反射池,穿过水平的入口,一眼就能瞥见这个通道。这是门廊入口设计的焦点,仿效Andy最喜爱的风箱式照相机设计而成。

室内入口呈长排形,穿过前厅映入眼帘的是层层叠叠的水景。水景可以为通道降温,也可以充当建筑内的一处绿洲和整个空间的幕景。这处住宅属于内向型设计风格。钢筋混凝土墙体构造起到了良好的保温效果,保证室内气温凉爽宜人。在整体结构内,建筑师通过结合不同的采光效果和绿化处理,精心打造优良的空间品质。

景观设计与建筑的结构设计融为一体。各种不同的植物/树种分别发挥着特定的作用——提供视觉线索、界定空间、连接空间、提升感官效果、有益健康、提供食材。房屋的侧立面由若干细长的垂直开口构成,这些开口按照房间的布局留出一定的间隔。这种处理方式为邻里之间保持必要的隐私性提供保障。屋顶设有阶梯式花园,尽可能地提供最优化的视觉效果——是极佳的野餐和观星场所。酷暑时节,屋顶的植物有效地发挥隔热效果,为邻里街坊提供凉爽的生活环境。

这栋别墅的代表性特点是体现了家庭。结实的门面隐藏着内部的开放与透明;坚硬的混凝土外壳之下是一处暖心的庇护之所;外观平平,却掩藏着持久凉爽宜人的室内环境,足以应对热带气候。这座住宅是一个家庭客户与建筑师之间的伟大合作的成果。在当代这样的一个热带气候地区,建造一所适合一家几代人共同居住、并能同时满足各人所需及愿望的住宅,该建筑便是用实例验证了这一构想的可行性。

Namly House

Peter and Lucy wanted a home where three generations could be housed under one roof, so that they can look after the grand-children, without having to compromise on freedom, differing needs and privacies of each generation. Peter is a fan of concrete architecture, and his brief was to "keep it clean, simple, and white". In addition, his wish list was "one house yet two homes, simple yet beautiful, front yet back, tropical yet cool and breezy, natural yet tasteful, raw yet elegant, indoor yet outdoor in touch with nature, small yet spacious, minimal yet more than enough, less yet more". The grandparents' room is located at the rear on the ground floor, easily accessible yet private, with Peter's deck next to it. This deck is a continuation of the dining-living space, to foster interactions amongst the generations.

The parents' room is located upfront on the second story, overlooking the twins' room separated by an air well. This configuration sets a balance in terms of privacies and harmonious living for the two generations.

The children's room is centrally located on the second story, between the parents' room and the library. This allows visual and

Namly别墅
Chang Architects

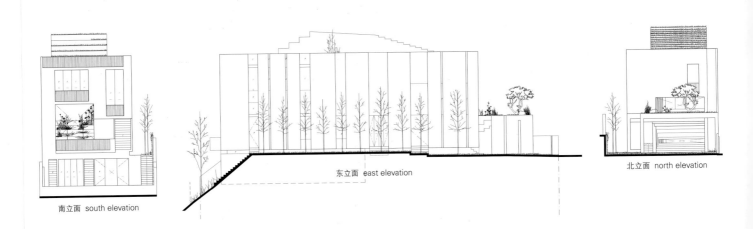

南立面 south elevation　　东立面 east elevation　　北立面 north elevation

physical access for the kids and parents. The space has the vantage of views and ventilation on both ends, and can be sub-divided if the twins so decide when they grow up.

Binding the spaces is a thoroughfare spanning front to the rear. This thoroughfare is the gathering space, the children's favorite playground, and effectively a wind tunnel for both the NE and SW monsoon winds. Upon arrival on car, one catches a glimpse of this thoroughfare, via a reflecting pond, through a horizontal aperture. This is the focal point of the entry porch, fashioned after the folding camera – Andy's favorite.

Entry to the interior is a procession via a vestibule, to be greeted by a cascading water feature, translated from the same fold. This water feature serves as a cooling agent for the thoroughfare, a green oasis and a backdrop for the spaces. This house is inward-looking. The interior, being heavily insulated by the reinforced concrete walls, is cool all the time. Within the monolithic structures, spatial qualities are crafted by varying daylight effects, integrated with plants.

Landscape is integral of the architectural design. Specific plant/tree species serve their respective roles as visual cues, space definers, connectors, for sensory-enhancements, for therapeutic purposes, and for food. The side elevation is a composition of slim vertical apertures, spaced intermittently to the room configurations. This is to accord due privacy for both neighbors. The house culminates with a roof terraced garden, optimizing the best possible view – a favorite spot for picnics and star gazing. During hot days, the planters are effective heat insulators, and contribute to a cooler environment for the neighborhood.

This house characterizes the family. Its solid front belies an interior that is open and transparent; beneath its concrete shell is a sanctuary of heart-warming dwelling spaces; and its stoic monolithic presence conceals a constantly cool environment that is responsive to the tropical climate. This house is a manifestation of great collaboration between the family and their architect. It demonstrates the potential of housing a multi-generation family serving differing needs and aspirations, in a contemporary tropical setting.

1. 别墅车道 2 泊车门廊 3 阶梯式水景 4 餐厅 5 起居室 6 储存室 7 客房 8 菜地
9 父母的房间 10 浴室 11 儿童房间 12 家庭娱乐室 13 阳台 14 安迪的工作室
15 午休室 16 机械/电子设备室 17 屋顶露天平台 18 阶梯式花园

1. Namly drive 2. car porch 3. cascading water feature 4. dining room 5. living room 6. storage 7. guest room
8. vegetable plot 9. parents' room 10. bath 11. children's room 12. family room 13. balcony 14. Andy's studio
15. nap room 16. mechanical/electrical service room 17. roof deck 18. terraced garden

A–A' 剖面图 section A - A'

1. alum. angle drip, powdercoated white
2. 40mm gap air-vent
3. clear, fixed, frameless tempered-laminated glass
4. 75 x 25 x 4.5mm thk hot-dipped galv. m.s. supports
5. fair-faced reinforced concrete wall
6. gypsum board false ceiling
7. built-in timber cabinets
8. internal wall, white putty finish
9. vertical sashless sliding window
10. timber floor / Balau floor planks, tanalized base, leveling screed, r.c. slab
11. metal mesh roller shutter box concealed within false ceiling
12. calcium silicate board false ceiling
13. concealed alum. guiding tracks for roller shutter
14. eco-pond / loose river pebbles, drainage cells, non-toxic epoxy sealer, protective screed, waterproofing membrane, r.c. slab
15. landscape planter through earth
16. terraced roof garden / planting medium, separation layer, drainage mat, protective screed to fall, waterproofing membrane, r.c. terrace
17. r.c. party / parapet wall in fair-faced reinforced concrete finish
18. Balau timber decking on timber batten
19. tempered laminated glass skylight with air-vents
20. air buffer, ventilation & light shaft
21. r.c. structural wall in fair-faced reinforced concrete finish
22. white putty finish
23. shower set
24. homo. tile finish to 1.8m ht waterproofing to 1.8m ht
25. shower floor / natural stone finish, protective screed to fall, waterproofing membrane, r.c. slab
26. steel C-channel termination for false ceiling
27. cascading water feature / lava rock finish, non-toxic epoxy sealer, protective screed, waterproofing membrane, r.c. slab
28. movable pre-cast concrete slab supported
29. s.s. plate forming pond wall, overflow cement screed finish
30. coral tree
31. landscape / planting medium, separation layer, drainage mat, protective screed, waterproofing membrane, r.c. slab to fall
32. integrated planter roof, car porch
33. 12mm thk, frameless, tempered laminated glass skylight
34. structural wall in fair-faced reinforced concrete finish
35. panel division lines
36. car-porch slab in fair-faced reinforced
37. sand binding on hardcore
38. pearl grass

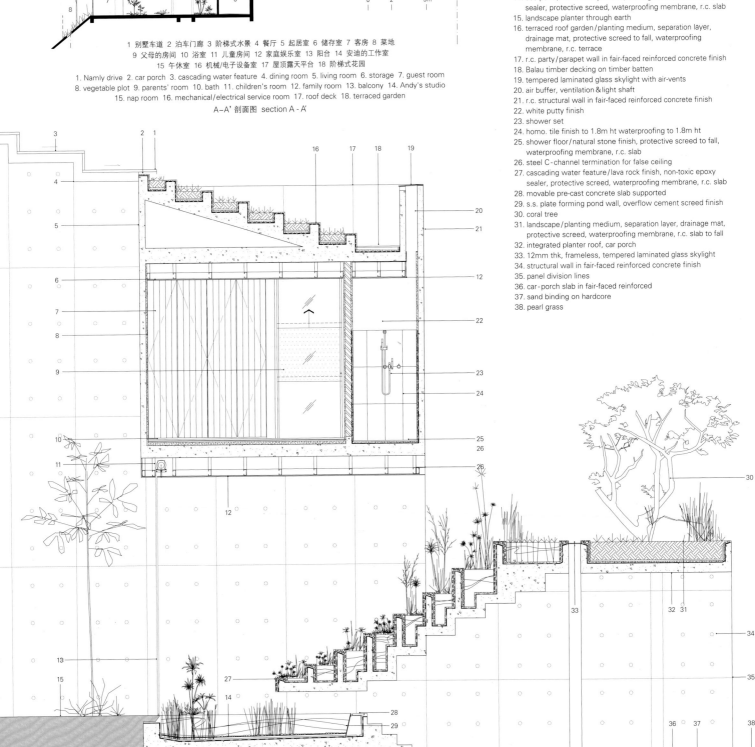

详图1 detail 1

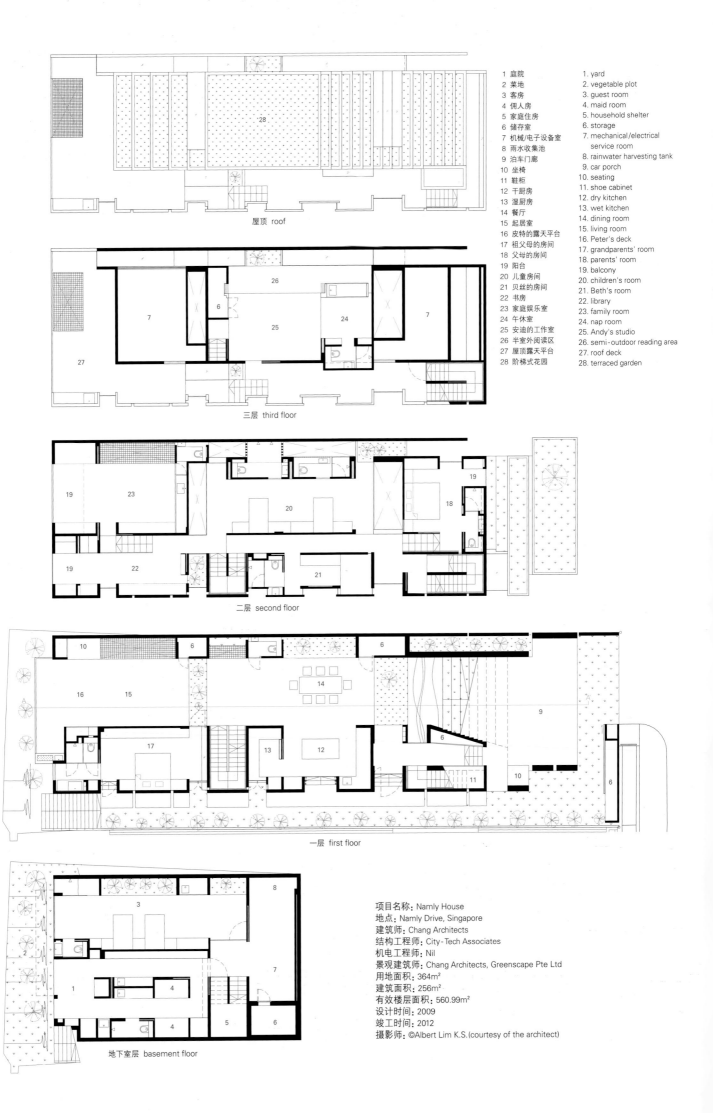

西原墙
Sabaoarch

新在宅:空间关系—地域性—隐私性 Urban Dwell: Proxemics–Territoriality–Privacy

"坚固并透明"的"Kabemado（开放式围墙）"能使人置身其中，并产生树影的效果。

位于东京的这座住宅地处两条路之间的3m宽的小地块内。建筑师在一处小型的间隔区内建造了一片居住空间，居民们可在此居住，十分安全。多个楼层由一串台阶连接，呈现出的效果像是沿着大树攀爬，伴随着周围的风景一直通往天际。

起初，建筑师将这面薄薄的坚固的墙想象成薄膜，在建筑、人类、物质和自然现象之间形成它的界限。

这种不平坦的结构交替形成雪松材质的纹理，使混凝土墙在粗糙的质地上形成阴影效果，好像带有陈旧缺口的墙体依存于周围带有丛林的居民区里。

为了划定使人们在两边有路的狭窄的空间里生活的范围，人们考虑了这种覆盖了主体的"Kabemado"（开放式围墙）结构，这个结构可与外界维系联系。其厚度使其可以充分地与外界隔离开来，形成一个内部修建的狭小的居所。

混凝土墙体上的小窗户给人一种深入内部和黑暗的感觉，这是由最初住所出现的树形成的。最后，建筑师开发了一处树形成的黑暗结构空间，并且这处空间成为一座试验性重建建筑。

由此形成了一个从外部感觉明显关闭的结构，同时居民又可与城市连接，并呈现绿植唤起的景象，内部是开放式的，实现原生态的、黑暗幽深的区域。

The Wall of Nishihara

"Strong and transparent "Kabemado(opening-wall)" makes people exist there and produces the darkness in trees."

This house in Tokyo stands at a small site with 3m width between two roads. The architects have created the space for dwelling in the small gap where residents could live in secure. The multi leveled floors are connected by rope of stairs and its experience is like climbing up the tree to the sky looking at the scenery of site around.

At first, the architects have visualized thin and strong wall as a membrane which contends on the border line among architecture, person, substance, and phenomenon.

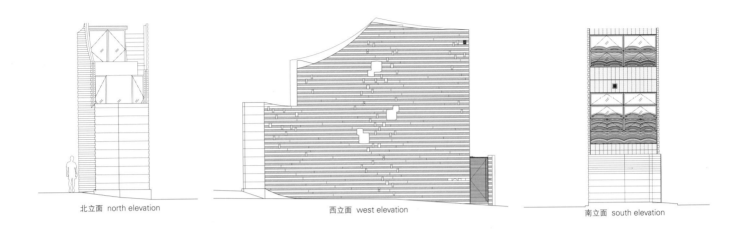

北立面 north elevation | 西立面 west elevation | 南立面 south elevation

项目名称：The Wall of Nishihara / 地点：Shibuya, Tokyo / 建筑师：Masanori Kuwabara
结构工程师：Matou Hayata / 承包商：NK / 用途：residence
场地面积：40.12m² / 总建筑面积：24.06m² / 有效楼层面积：78.06m² / 结构：reinforced concrete / 竣工时间：2013.12
摄影师：©Ohno Shigeru (courtesy of the architect) - p.151, p.153bottom, p.155
©Yuji Nishijima (courtesy of the architect) - p.148~149, p.150, p.152, p.153top, p.154

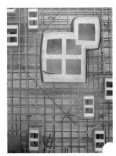

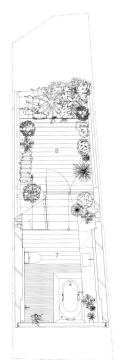

四至五层
third-fourth floor

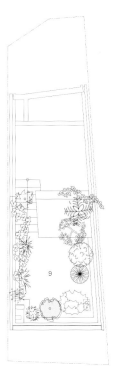

屋顶
roof

1 入口
2 房间2
3 房间3
4 房间4
5 露台
6 厨房
7 浴室
8 屋顶1
9 屋顶2

1. entrance
2. room 2
3. room 3
4. room 4
5. patio
6. kitchen
7. bathroom
8. roof 1
9. roof 2

一至二层
ground-first floor

三至四层
second-third floor

The unevenness made in laps turns into a cedar mold, and a concrete wall creates the shade on the rough texture. It seems that the wall with the worn opening is parasitic on the residence image with the hedge forest around the site.

In order to make the domain in which man can live in the narrow place with a both-sides road, the detail of "Kabemado" which wraps the body was able to be considered, maintaining relationship with outside. The rich depth which may fully feel distance with the exterior as an abode also into thinness is produced inside. The small window dug on the concrete wall brings feeling of "Okusei(inner depth)" and "Yami(darkness)" which trees as the origin of a dwelling make. Finally, they developed once the structure of space with the darkness which trees have, and became a trial which is reconstructed as an architecture.

As a result, while the exterior apparently felt closed space, it connects habitants with a city, and also provides an image as the whole trees evoke. The inside was open bringing about rawness with the darkness of space, and the depth. Sabaoarch

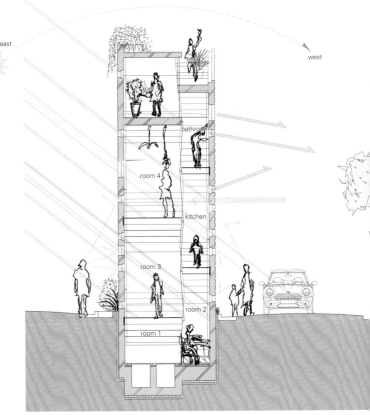

A-A' 剖面图 section A-A'

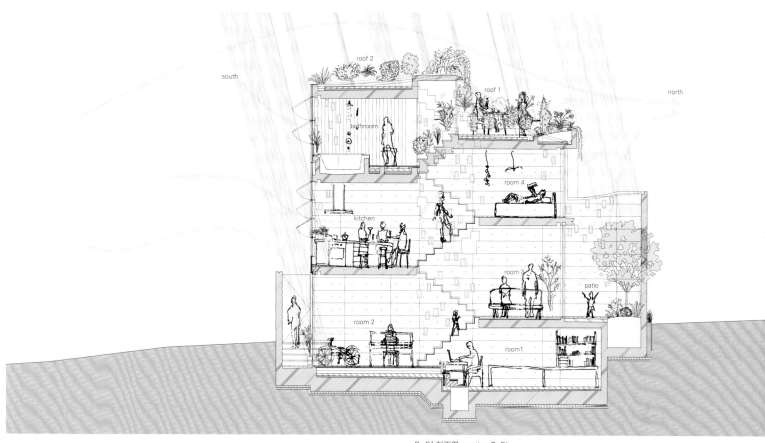

B-B' 剖面图 section B-B'

津田沼住宅
Fuse-Atelier

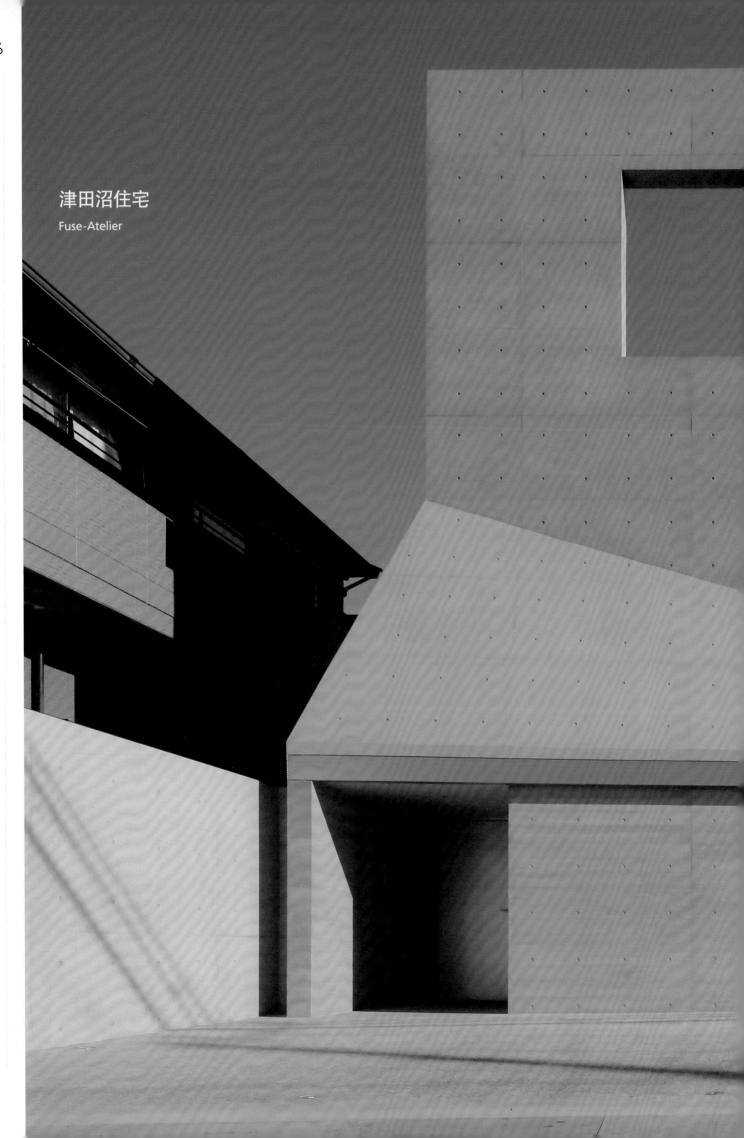

该住宅位于奈良市,是为一对四十多岁的夫妇建造的住所。房子地处住宅区和商业区的边缘,商业区从车站开始延伸,且附近有一个大型商场。房屋面向一条8m宽的大街,大街属于城市道路,因邻近火车站,所以交通非常繁忙。客人希望采用混凝土来建造房屋,那样所有的空间可以按顺序连接起来,同时也可以保留一处隐私的、宽敞的生活空间。

由于邻近4m宽的公路属于城市规划区的一部分,所以建筑师在整个场地的后面建造房屋。由此该结构前面的空地可以作为街道和住宅之间的缓冲区。考虑到隐私性和外面的噪音,建筑师将一处户外空间融入到结构当中。由于不同的法律要求,建筑结构的上部需要向后移。因此,结构的外部逐渐由两层变为三层。

该住宅内的不同空间通过设计尽可能多的视线和流动线条而连接起来。

带有一条通道(从二楼阳台一直延伸至屋顶)的垂直嵌入的室外空间有效地隔离了内部的空间。这种做法结合了内部的流动性规划,也扩大了建筑的整体导向效果。

除了内部和外部的流动性规划,建筑师分层设置了视线网来激发不同空间的联系效果。相关联部分的堆叠在水平和垂直两个方向同时形成了缺口,使人们通过令人惊奇的躲避方式看到各种顺序的复杂层次。通过完全融合混凝土、玻璃、金属和岩石的极简抽象艺术的细节,使空间增强了明显的明锐度,立体结构的框架也变得容易看见。这样房子成为一个动态连接的、三维立体的空间链,从而激发和产生了各种各样的效果和距离。

House in Tsudanuma

The house is for a married couple in their forties, located in the city of Narashino. It is situated on the edge of a residential neighborhood and a commercial district that stretches from the station, with large-scale stores nearby. The house faces an eight meter wide street classified as an urban roadway, with busy traffic due to its proximity to the train station. The client wished for a concrete house, where all spaces are stringed together in sequence, while also maintaining the privacy and freedom of a spacious living space.

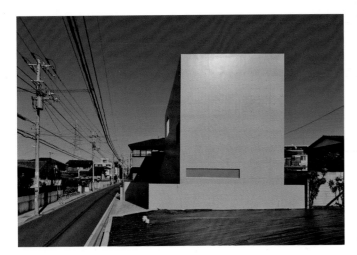

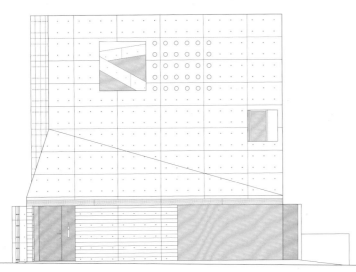

西立面 west elevation

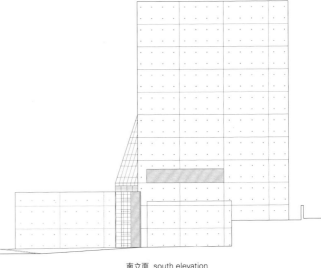

南立面 south elevation

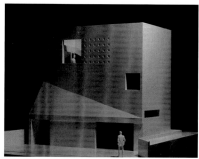

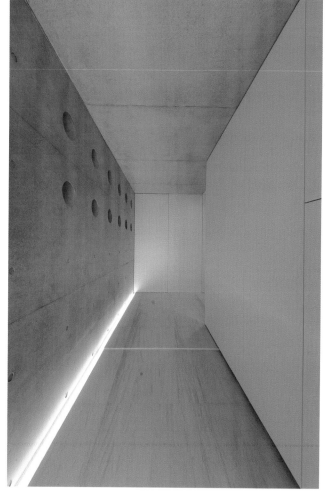
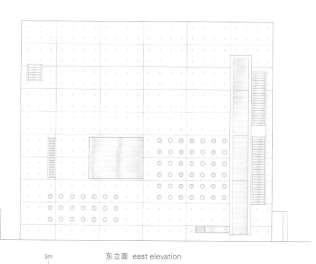

东立面 east elevation

项目名称：House in Tsudanuma / 地点：Narashino, Chiba Pref., Japan
建筑师：Shigeru Fuse / 设计团队：Fuse-Atelier + Musashino Art University, Fuse-Studio
结构工程师：Ysutaka Konishi / 主要承包商：Three F / 竣工时间：2014
结构：reinforced concrete / 用地面积：133.27m² / 总建筑面积：63.83m² / 有效楼层面积：153.41m² / 材料：exposed concrete, natural genuine stone
摄影师：courtesy of the architect

162

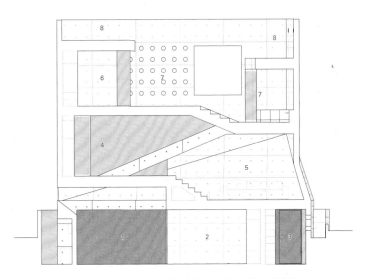

1 车库 2 工作室 3 入口 4 厨房 5 起居室 6 卧室 7 露台 8 屋顶露台
1. garage 2. studio 3. entrance 4. kitchen 5. living room 6. bedroom 7. terrace 8. roof terrace
A-A' 剖面图 section A-A'

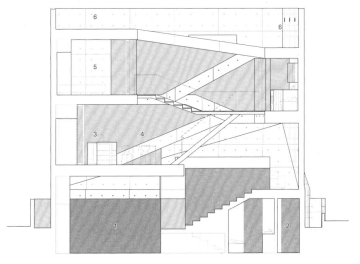

1 车库 2 入口 3 厨房 4 餐厅 5 卧室 6 屋顶露台
1. garage 2. entrance 3. kitchen 4. dining room 5. bedroom 6. roof terrace
B-B' 剖面图 section B-B'

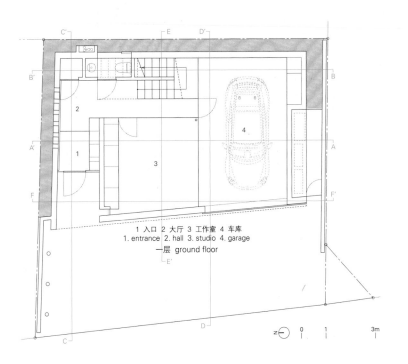

1 入口 2 大厅 3 工作室 4 车库
1. entrance 2. hall 3. studio 4. garage
一层 ground floor

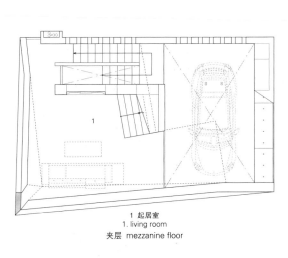

1 起居室
1. living room
夹层 mezzanine floor

Since four meters of the adjacent road is specified as part of urban zoning, the architects built the structure in the back part of the site. The resulting open space in the front of the structure functions as a buffer between the street and the house. Concerned with the privacy and noise from the outside, the architects incorporated an outdoor space into the structure. Upper parts of the building structure had to be set back due to various legal limitations. As a result, the exterior of the structure was tapered from the second to third floor.

The various spaces of this house are interconnected by producing as many lines of sight and flow as possible.

The vertically inserted outdoors space with a passage that leads from the second floor terrace to the rooftop effectively breaks up the interior space. Combined with the flow planning of the interior, this also amplifies the building's overall ease of navigation.

In addition to the flow planning on the exterior and interior, the architects layered a network of sight lines to inspire diverse spatial relationships. An accumulation of associated parts creates gaps both horizontally and vertically, leading the eye through surprising escapes that result in a complex layer of various sequences. By fully embracing the minimalist details of concrete, glass, metal and rock, the space is intensified with distinct sharpness, where the skeletons of the spatial structure become visibly apparent. The house becomes a dynamically linked, three-dimensional chain of spaces that inspires various scenarios and distances.

Shigeru Fuse

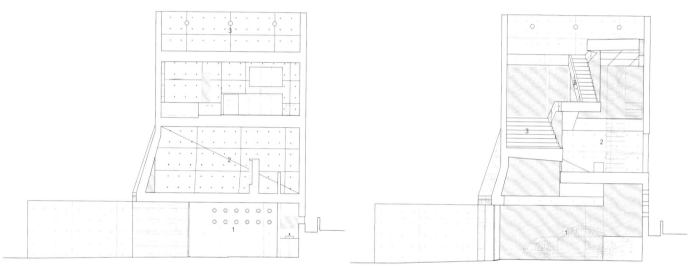

1 入口 2 起居室 3 屋顶露台
1. entrance 2. living room 3. roof terrace
C-C' 剖面图 section C-C'

1 车库 2 厨房 3 露台
1. garage 2. kitchen 3. terrace
D-D' 剖面图 section D-D'

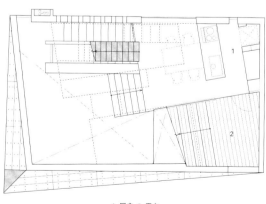

1 厨房 2 露台
1. kitchen 2. terrace
二层 second floor

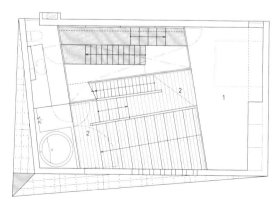

1 卧室 2 露台
1. bedroom 2. terrace
三层 third floor

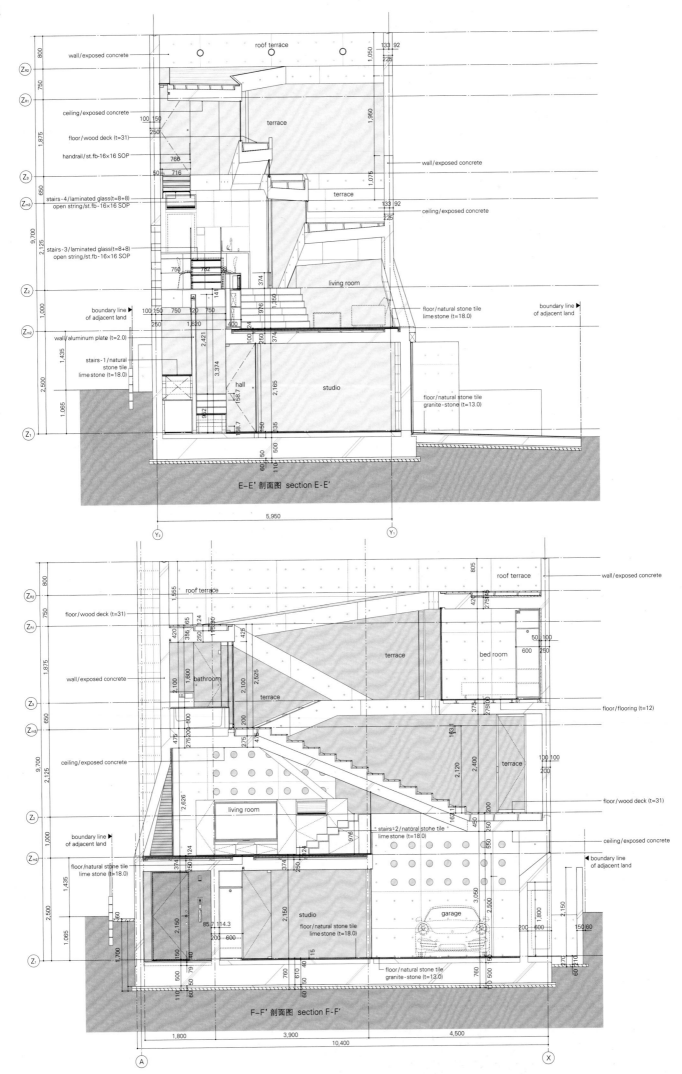

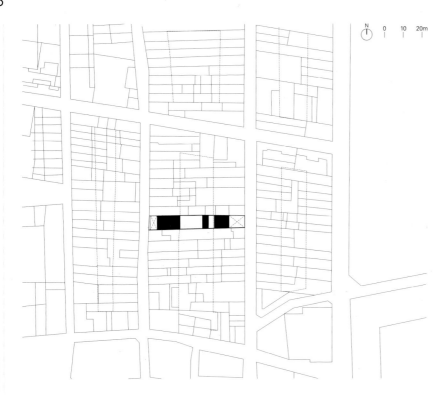

1014住宅

Harquitectes

场地地处历史悠久的格拉诺列尔斯市中心，融入了住宅界墙之间的城市肌理中。可用的空间狭窄而细长，只有6.5m宽，人们可从两条街道进入其中。现有的建筑破旧不堪，只有主立面具有一定的遗产价值而保存完好，从而使建筑得以保留。

业主的愿望是将住宅划分为两处不同的区域：一处是用于日常生活的家庭空间；另一处则是安静的空间，与前者隔绝，可以独立用作客厅，来欢迎偶尔来访的客人，或者组织聚餐和聚会。家庭空间面向主街道而建，而场地西侧被中央花园分隔开，成为一处单独的区域，供车辆进出。

由于场地是东西朝向，加上通往街道的通道狭窄，导致建筑很难通过面向街道的立面来较好地吸收日光。加上一层的隐私性较差，使建筑师从前侧的街道线开始将建筑向后移，并且在房子两侧形成入口天井。这些天井在获取上方的日光的同时也在街道和房屋、室内和室外气候之间创造了过渡空间。带有可伸缩屋顶的半覆盖的实用空间能够在冬天采集能源，在夏天进行通风。这种方法解决了行人从主干路进出建筑以及另一侧汽车的通行问题，避免了这种用途产生的典型的二次空间及经常被忽视掉的空间。这些空间所拥有的隐私、照明和热舒适度使住宅的每个区域都得以利用，没有任何隐藏的或者多余的空间。这些生物气候空间成为建造连接街道与其他设施的一系列空间的第一步。这一系列的空间和热条件形成了一个53m长、总面积为345m²的底层。同时，这些空间还作为一个连续的长廊，分别通往上层的私人区域与服务区域以及地下室。

住宅内的每个房间都进行了个性化的处理，并且彼此之间相互连接，以清晰地确定作为整体的一部分的每处空间的特定用途。这种做法使外部空间产生了和起居空间一样的效果，变成住宅内的另一个房间。这样，底层将不同的室内空间、错落有致的屋顶、狭长形的半室外覆顶房间（符合生物气候学），以及覆顶和露天的室外房间结合起来。

House 1014

The plot is located in the historical city center of Granollers and placed into an urban fabric of dwellings between party walls. The available space is narrow and elongated, only 6.5m wide, and accessible from two streets. Owing to the dilapidated state of the existing building, only the main facade, reasonably well preserved offering a certain heritage value, could be maintained.

The desire of the owners was to differentiate two zones in the house: the domestic area, where the daily family life had to take place; and a secondary area, quiet and isolated from the first, functionally independent to be used indistinctly as a living room or to welcome occasional guests as well as organize meals and gatherings. The domestic area is placed towards the main street, while the west side of the plot, isolated and separated by the central garden, becomes a segregated zone for vehicle access.

The east-west orientation of the plot and the narrowness of the access streets made it difficult to achieve a good solar gain throughout the street facades. This circumstance, added to the difficulties to achieve privacy on the ground floor, led the architects to move back the building from the front street line, creating access patios on both sides of the house. These patios allow upper solar gain and at the same time create transition spaces, both physically between the street and the house, and between outdoor and indoor

三层 second floor

二层 first floor

一层 ground floor

地下一层 first floor below ground

1 卧室 2 浴室 3 工作室 4 天井 5 厨房 6 起居室 7 烧烤区 8 集会室 9 机械间 10 存储室 11 服务室 12 洗衣房 13 客房
1. bedroom 2. bathroom 3. studio 4. patio 5. kitchen 6. living room 7. barbecue place 8. gathering room 9. machine room 10. storage 11. service bedroom 12. laundry 13. guest room

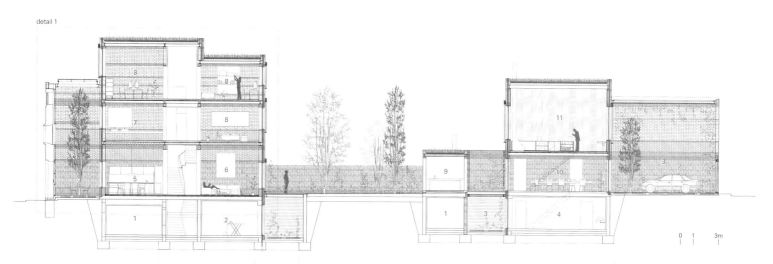

1 存储室 2 洗衣房 3 天井 4 客房 5 厨房 6 起居室 7 浴室 8 卧室 9 烧烤区 10 集会室 12 工作室
1. storage 2. laundry 3. patio 4. guest room 5. kitchen 6. living room 7. bathroom 8. bedroom 9. barbecue place 10. gathering room 11. studio
A-A' 剖面图 section A-A'

巴塞罗那街道一侧的立面
facing Barcelona street

B-B' 立面图_天井一侧视野
elevation B-B'_a view from patio

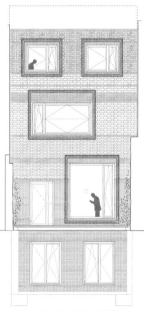

C-C' 立面图_花园一侧视野
elevation C-C'_a view from garden

Ricomà街道一侧的立面
facing Ricomà street

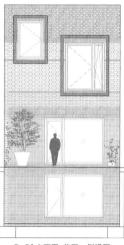

D-D' 立面图_花园一侧视野
elevation D-D'_a view from garden

E-E' 立面图_天井一侧视野
elevation E-E'_a view from patio

0 1 2m

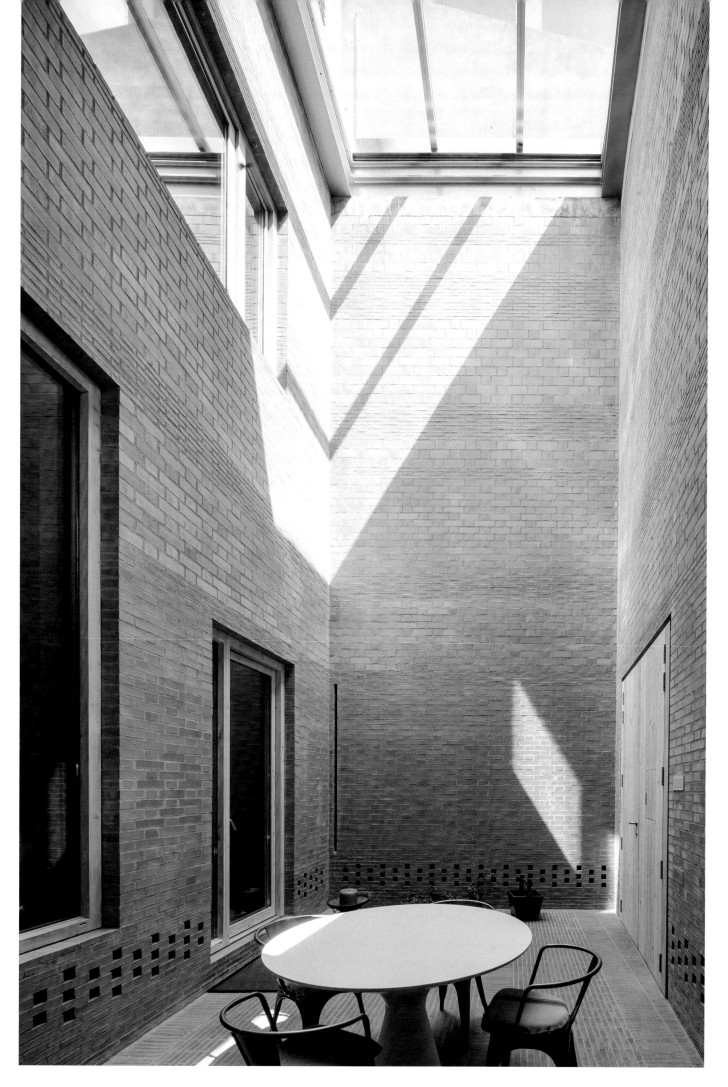

climate. Semi covered and practicable spaces with retractable roof tops allow capturing energy during winter time as well as ventilation through summer. This way both, pedestrian entrance from main central street and vehicles access from the opposite side are solved, avoiding typical secondary and often neglected spaces that are often generated by this kind of uses. The qualities of privacy, light, space and thermal comfort of these spaces allow the house to be used and perceived from end to end, without any hidden or residual space. These bio-climatic spaces become the first step of a sequence of areas linking one street to the other offering. The addition of this sequence of spaces and thermal conditions creates a ground floor 53m long totaling 345m². It works at the same time as a long continuous hallway, giving access to the private and service areas of the house located on the upper floor and the basement, respectively.

Each one of the rooms was individually treated yet carefully connected with the next one, clearly identifying the specific use of each space being all, at the same time, part of a whole. This approach helps exterior spaces to achieve the quality of a living space, therefore becoming one more room of the house. This way, the main ground floor combines different interior rooms, lower or higher roofs, longer, semi-exterior covered and bio-climatic rooms as well as covered and uncovered outdoor rooms.

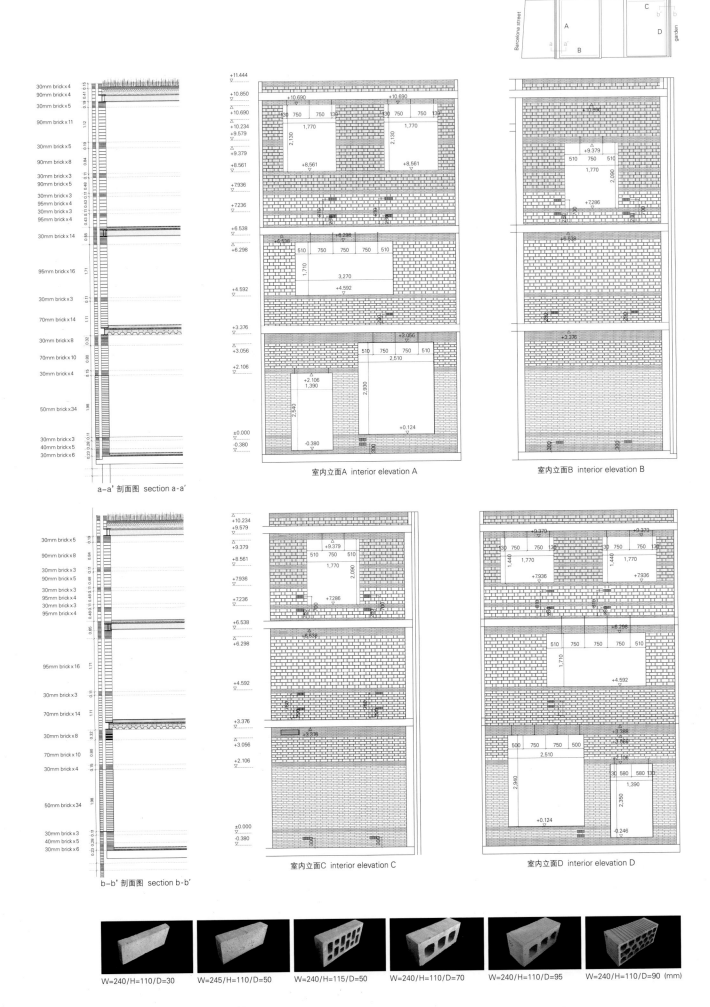

项目名称：House 1014 / 地点：Granollers, Barcelona, Spain
建筑师：David Lorente, Josep Ricart, Xavier Ros, Roger Tudó
合作：Blai Cabrero Bosch, Montse Fornés Guàrdia
工料测量师：Carla Piñol Moreno, Ramon Anton
室内设计师：Fátima Vilaseca
结构工程师：DSM Arquitectes
安装：Igetech / Àbac enginyers
景观建筑师：Anna Esteve
用地面积：364.96m² / 总建筑面积：236.75m² / 有效楼层面积：672.90m²
设计时间：2010.11 / 施工时间：2012.14 / 竣工时间：2014
摄影师：©Adrià Goula (courtesy of the architect)

©Thomas Grimes

>>140

Chang Architects
Chang Yong Ter was born and raised in Singapore, Yong Ter's passion for architecture was discovered during his university years at the School of Architecture, National University of Singapore. Upon graduation, he sought apprenticeship with Mr. Tang Guan Bee for several years, before starting his practice, Chang Architects, at the turn of this millennium.
Yong Ter believes that architectural design is a work from the mind and the heart. While rationality and logic could fulfill functional briefs and achieve pragmatic efficiencies, an intuitive, poetic approach could resonate with the soul, and transcend limitations of rationalities. Therefore, part of the design process also includes unlearning and forgetting, and self-discoveries of the basics/origin.

>>74

Smith Vigeant Architectes
Daniel Smith[right] and Stéphan Vigeant[left] have been working together since 1992 to define an architectural practice that transcends the traditional scope of the profession. Their vision, combined with diverse and extensive professional experience, now extends to a wide range of projects from the residential, institutional, cultural, corporate, and urban design sectors. Their integrated and sustainable approach enables the team to pioneer solid and unique design solutions that effectively incorporate their collective expertise. With this holistic approach to design, all components of a building are integrated from the initial design phase until completion. The team at Smith Vigeant have built a deep portfolio that includes a broad array of projects varying in scope and complexity, that are cost-effective, energy efficient and have low environmental impact.

>>54

Tod Williams Billie Tsien Architects
Tod Williams[right] was born in Detroit, Michigan and received his undergraduate degree and Master of Fine Arts and Architecture from Princeton University. Billie Tsien[left] was born in Ithaca, New York and received her undergraduate degree in Fine Arts from Yale and her Master in Architecture from UCLA. Williams and Tsien began working together in 1977 and nine years later established their partnership, Tod Williams Billie Tsien Architects, in a ground floor space on Central Park South where they still work today. Most recently, they were awarded an honorary international fellowship to the Royal Institute of British Architects, the National Medal of Arts from the United States government, and the Firm Award from the American Institute of Architects.

Aldo Vanini
Practices in the fields of architecture and planning. Had many of his works published in various qualified international magazines. Is a member of regional and local government boards, involved in architectural and planning researches. One of his most important research interests is the conversion of abandoned mining sites in Sardini.

Fabrizio Aimar
Graduated cum laude from Politecnico di Torino(Italy). Worked for five years for a civil and infrastructural engineering office in Torino. During this period, he has developed different structural projects in collaboration with some international architectural firms such as Jean Nouvel, Renzo Piano, Mario Cucinella and Aymeric Zublena. In 2014, Fabrizio has worked on the new Intesa Sanpaolo skyscraper in Torino designed by Renzo Piano(RPBW), and in the same year he has founded his own practice. Since 2010, he has been a contributor for the Italian architectural magazine "Il Giornale dell'Architettura" and also a member of Cultural Committee of the Asti local board of Architects. Since 2014, he is an external writer for the Italian technical websites "architetto.info" and "ingegneri.info".

>>46
Kiss + Cathcart
Is located in the Brooklyn, New York founded by two parters; Gregory Kiss and Colin Cathcart in 1983. Registered Architect of NY, Gregory Kiss received Bachelor of Arts from the Yale University in 1979 and Master of Architecture from Columbia University in 1983. Received several honors from the AIA Committee. Also won the First Prize at the 1996 Photovoltaics in Buildings competition, 1996 AIA BIPV Competition and 1992 Solar Parking Structure Competition.

>>64
Doazan + Hirschberger & Associés
Is a group of eight to ten people composed of architects, urban planners, landscape architects and administrative staff. Was founded in France, 1998 by Benoîte Doazan[third] and Stéphane Hirschberger[fourth]. In 2010, Nicolas Novello[second] and Elisabeth Salvado[first] joined as associate architects of the firm. Benoîte Doazan was born in 1967 and obtained Architect DPLG in 1992. Stephane Hirschberger was born in 1966 and received Architect DPLG in 1992. Has been a teacher since 2004 at the Superior National School of Architecture and Landscape of Bordeaux(ensapBx).

>>132
Bruno Vanbesien + Christophe Meersman
Bruno Vanbesien was born in Aalst, Belgium in 1976 and graduate from the Luca School of Arts(LUCA) in Brussels in 2001. From 2002 to 2005, he worked at Pascal François Architects and established his own office in 2005. Christophe Meersman was born in Vilvoorde, Belgium in 1977. Worked as an collaborator at Govaert-Vanhoutte Architects and D+A Consult Architects office.

Harquitectes

Ia an architecture studio established in Sabadell, Barcelona, 2002. Has been operated by four partners; David Lorente Ibáñez[first], Josep Ricart Ulldemolins[second], Xavier Ros Majó[fourth] and Roger Tudó Galí[third]. All of them studied at Vallès Higher Technical School of Architecture from 1998 to 2000. Their work have been recognized with several awards and also selected for local and abroad exhibitions.

>>148

Sabaoarch

Masanori Kuwabara is a member of the Architectural Institute of Japan(AIJ) and Japan Institute of Architects(JIA). Was born in Gifu, Japan in 1973. Received a B.Arch in 1996 and M.Arch in 1998 from the Faculty of Architecture at Waseda University. Has worked at Atsushi Kitagawara Architects for 8 years before founding Sabaoarch in 2008. He was also a lecturer of his alma mater, Waseda University from 2011 to 2014.

>>34

CEBRA

Lars Gylling was born in 1973 and graduated from the Aarhus School of Architecture in 2000. Was formerly employed at 3XN(2000~2006). Is an versatile all-round architect with extensive experience and insight into all aspects and phases of the profession. Approaches every project in high spirits and lives by the motto "It is yourself who makes work exciting". Mikkel Frost was born in 1971 and graduated from the Aarhus School of Architecture in 1996. Is a member of the Royal Institute of British Architects(RIBA). Is part of the daily management of the office and is naturally part of the business affairs. Also regularly lectures at architecture schools and events around the world.

>>84

Lanz+Mutschlechner

Martin Mutschlechner[first] established architecture studio Stadtlabor.org in 2001, based in Toblach, Italy and Innsbruck, Austria. Babara Lanz[second] is working as freelance designer and art historian, mainly in the historic building research. They operates extensively and successfully across a broad range of architectural disciplines in inner city, urban, provincial and rural environments.

Glifberg+Lykke

Rune Glifberg is one of the world's best skateboarders. Therefore he has seen and skated the world's best skate parks for 20 years. He has devoted a large amount of his time into designing and creating skate parks in Denmark. Ebbe Lykke studied Design and Architecture at the Royal Danish Academy of Fine Arts. Received a Master in Design, from Karch(2009) and has been managing his own design company since 2003. In 2009, as old friends with common interests, Glifberg and Lykke started designing urban spaces together in connection with the Multipark in Elsinore, Denmark.

Wolfgang Meraner

Wolfgang Meraner[third] studied architecture at the University of Innsbruck and gained experience at Gostner, Kerschbaumer, and Pichler and Partner. In 1999, he opened his own architectural office.

ISON Architects

Jean Son, one of the founding partners of ISON Architects, studied at the Venice Institute of Architecture(IUAV) after receiving a B.Arch from the Hongik University, Seoul in 1986. Has worked at the Studio Francesco Venezia, Naples and Studio Skopje, Macedonia before establishing his own office. Was a Adjunct professor of the Hongik University, Korea University and the Korean National University. Has been teaching at the Graduate School of Architecture in KonKuk University since 2014. Received KIA Award twice from the Korean Institute of Architects and Kim Swoo Geun Prize from the Kim Swoo Geun Goverment in 2008.

>>156
Fuse-Atelier

Shigeru Fuse was born in Chiba, Japan in 1960 and graduated from the Department of Architecture, Musashino Art University. Between 1985 and 2002, he worked at Daiichi-Kobo Associates and in 2003, he established his own atelier, Fuse-Atelier. Now he is a professor at Musashino Art University.

Nextoffice

Alireza Taghaboni was born in Tehran, Iran in 1977. Commenced his university studies in 1995 at Gilan University, in the field of Architecture. Upon receiving his master degree in Architecture in 2002, he began a Ph.D. program at "Azad University of Science and Research of Tehran" and graduated in 2007. After broad experiences of 8 years in different architecture offices, he began his own practice in 2004 and established Nextoffice in 2009. Designed more than 50 projects in the past 10 years and has received numerous architecture awards, including first and second prize of "Memar Award". Furthermore, he had been teaching at different universities since 2002.

>>112
Tuttiarchitetti

Vincenzo Giusti(1961), Daniela Finocchiaro(1963) and Luigi Pellegrino(1963)from the left have been working in the field of architecture more than 20 years. After some experiences in common, they have joined by the name of Tuttiarchitetti in 2011. The components of the study have won awards and prizes by participating in various national and international competi-

C3, Issue 2015.6

All Rights Reserved. Authorized translation from the Korean-English language edition published by C3 Publishing Co., Seoul.

© 2016大连理工大学出版社
著作权合同登记06-2016年第21号
版权所有·侵权必究

图书在版编目(CIP)数据

城市复兴中的生活设施：汉英对照 / 韩国C3出版公司编；史虹涛等译. — 大连：大连理工大学出版社，2016.4
 (C3建筑立场系列丛书)
 书名原文：C3：Amenity in Urban Revival
 ISBN 978-7-5685-0340-2

Ⅰ. ①城… Ⅱ. ①韩… ②史… Ⅲ. ①城市公用设施－建筑设计－汉、英 Ⅳ. ①TU984

中国版本图书馆CIP数据核字(2016)第058006号

出版发行：大连理工大学出版社
　　　　　（地址：大连市软件园路80号　邮编：116023）
印　　刷：上海锦良印刷厂
幅面尺寸：225mm×300mm
印　　张：11.75
出版时间：2016年4月第1版
印刷时间：2016年4月第1次印刷
出 版 人：金英伟
统　　筹：房　磊
责任编辑：许建宁
封面设计：王志峰
责任校对：王　伟
书　　号：978-7-5685-0340-2
定　　价：228.00元

发　行：0411-84708842
传　真：0411-84701466
E-mail：12282980@qq.com
URL：http://www.dutp.cn